SEXTECH REVOLUTION

(SEX) TECH

REVOLUTION

The Future of Sexual Wellness

ANDREA BARRICA

LIONCREST
PUBLISHING

SEXTECH REVOLUTION

The Future of Sexual Wellness

ISBN   978-1-5445-0493-3 *Hardcover*

978-1-5445-0491-9 *Paperback*

978-1-5445-0492-6 *Ebook*

978-1-5445-0639-5 *Audiobook*

For my Lolas, Pacita and Mila

CONTENTS

INTRODUCTION

When I pitch investors, I place a 3D-printed model of this on the table and ask:

Do you know what this is?

Can you recognize the structure? I'll give you some hints: It's an organ in the human body. It's inside roughly half of the population. It is densely distributed with nerve endings, and its only job is to experience pleasure. In fact, it's the most powerful pleasure organ in the human body.

It is the clitoris.

Did you recognize it? Don't worry. Most people don't.

None of the venture capital investors I've shown it to have had any idea what it is. Neither have many of the medical doctors I've met. (Perhaps that's because most medical students receive fewer than ten hours of sex education during their entire four years in medical school.[1])

Until relatively recently, the clitoris and other major aspects of human sexuality were largely ignored by the scientific community. The 1948 edition of Grey's Anatomy went so far as to omit the clitoris completely, but even today information on pleasure, especially female pleasure, is barely touched on in medical textbooks.[2] For years, the clitoris was regarded as having no reproductive role whatsoever, and in the nineteenth century, doctors even recommended removing it to prevent "hysteria."

When the clitoris is discussed today, it's talked about as a 1-2cm external tip, rather than the fully formed, comprehensive 10cm organ pictured above. It's thanks to researchers like Dr. Helen O'Connell, who produced one of the first fully realized anatomical depictions of

---

1   DS Solursh, "The human sexuality education of physicians in North American medical schools." *Int J Impot Res.* 2003 Oct;15 Suppl 5:S41-5.

2   Ibid

the clitoris in 1998, that we have a sense of its true scope and importance.

As HuffPo's Cliteracy project so aptly observed, we put a man on the moon in 1969, invented the internet in 1982, and didn't fully understand the anatomy of the clitoris until 1998.[3]

That's some seriously powerful stigma.

People don't like to talk about the concept of a powerful organ that solely exists to experience pleasure. The concept of pleasure itself is difficult for people to talk about—particularly with family, partners, or medical professionals.

Most of us are taught that sex is shameful, that pleasure is indulgent, and that using your body for one of its naturally designed purposes is somehow abuse. As a result, parents don't talk with children. Lovers don't communicate with partners. And medical providers often ignore the subject altogether.

Why is sex not a basic, normal part of wellness—and why has over half of the population been ignored?

---

3 Carina Kolodny and Amber Genuske, "Cliteracy," *Huffington Post*, May 2015.

SO MANY QUESTIONS

As an entrepreneur and the owner of a clitoris, this baffling question pushed me into the industry known as "sextech"—technologies, products, and services that seek to innovate and improve the human sexual experience.

There are 7.7 billion people on earth, and with the exception of babies born through in vitro fertilization and other procedures, each exists because two people had sex. We talk about sex all the time—in magazines, blogs, television series, jokes, sermons, movies, ads, fashion, podcasts, and porn. We're obsessed with sex, but who are the major brands that shape our daily experiences with sexuality? Where are the reputable, trusted voices?

As I researched this question, I found that for all our fascination with the subject, there are very few that provide an answer—and the ones that do tend to sensationalize it, rather than strip it down. Sexual wellness is like a huge meadow—a vibrant ecosystem carpeted in small plants and grasses, but no major trees.

Two years after launching O.school, and countless venture capital meetings, speaking panels, product development meetings, interviews and conversations with activists, entrepreneurs, doctors, and nonprofits in the space, I'm still learning. But I've realized that just as sex has no central place for information, neither does sextech. So many of us are working independently, siloed in our own corner, with few networking events or shared resources.

I'm writing this book for people who have the desire to push forward sexual wellness. I want to share what I've learned in my conversations with investors, founders, activists, policy makers, and educators.

Why is a market so large so behind in innovation, tools, resources, and solutions? Why are there so few places to talk about sex honestly and accurately—offline or online? What are the hurdles that have stopped the previous generation, and how can we begin to get around them? How can so many people have so many progressive ideas about sex and sexuality—and still fall short of creating mass change in our culture? How can the need be so great, and yet the market be so unfulfilled?

In short, what's the future of sexual wellness?

In this book, I'll talk about the growing sexual wellness industry, the problems, and challenges in the market for both customers and entrepreneurs. I'll lay out both the progress we're making and the roadblocks we face. I'll tell you about the trends that augur well for us, and where the market might take us. I'll lay out my vision of what the future of sexual wellness looks like, and some of the challenges we will need to overcome to get there.

I think I echo the sentiments of many of the leaders in the sexual wellness space who want it to be much bigger. Those of us in this space seem to understand that if, together, we can begin dismantling stigma and improving access, we all stand to benefit—as entrepreneurs and as a world community.

THE REVOLUTION IS HAPPENING

I am but one of a number of innovators in sexual wellness, from reproductive rights and pleasure activists, sex educators, feminist writers, sexologists and researchers, nonprofit innovators, sextech founders, and countless beacons of sex positivity serving their friends and communities with peer-led support and education. It's on their shoulders that I stand.

There are other books about the recent developments in sexual wellness worth calling out specifically, like Emily Nagoski's groundbreaking book, *Come as You Are*, one of the best modern books about the science of sexuality and desire. Esther Perel's book *State of Affairs* has shifted the way we think about infidelity, intimacy, and desire in modern relationships. Sonya Renee Taylor's *The Body Is Not an Apology* is a gem in the body positivity movement. Lynn Comella's *Vibrator Nation* covers the history of sex toy education and feminist sex stores. Adrienne Maree Brown's *Pleasure Activism: The Politics of Feeling Good* is an important read in the intersection of pleasure and activism.

Anyone working in this space needs to understand that sextech is like the clitoris itself—the visible part, the part getting all the attention, is actually just a small part of a much larger structure. To succeed, entrepreneurs and advocates in this space need to understand the contours of human sexuality, to understand the politics of inclusion and the history of this movement. While I can share my experiences and what I've learned, I'm at best a cluster of nerve endings.

In fact, one of the things we will cover is the evolving use of language around gender and health. Just as I've learned to code-switch my use of language between speaking with my parents in our Filipino home and speaking to people differently elsewhere, and using venture capital lingo and translating these terms to other communities, I've chosen to code switch sometimes between mainstream binary terms for gender and the new, inclusive language used in sexual wellness spaces. This means sometimes referring to industries like "femtech" and speaking in terms of "men and women." I will also sometimes use the new language of inclusive sexual wellness, which you may see when we talk about "people with vulvas" and "reproductive health." In chapter 5, we will dive deeper into why this is important.

This book is ultimately born of my experiences, and I'll focus on my expertise—accessing capital and power. Within that, I can talk about the gatekeepers who control it, and what they have taught me. I'll talk about the structural issues that have hindered our sexual wellness, from the politics of education to the foundation of the modern internet. And I'll share what I've learned, in hopes it can help others.

I'm betting on a world where sexual wellness is a thriving market for a healthier generation. I'm betting on a sextech revolution.

What's your money on?

{ CHAPTER 1 }

A BRIEF HISTORY OF THE INDUSTRY

The category of sextech is not entirely new, though it is making new strides. Make a visit to the Antique Vibrator Museum in San Francisco, and you'll see plenty of early attempts in the field. Later on, sex was used to push forward technology from the early film cameras to the VCR to the internet to VR—thanks to the driving force of porn.

But as an investment category, as a market sector worthy of innovation, the idea of sextech is quite new.

Entrepreneur Cindy Gallop is credited with coining the term sextech, following her groundbreaking 2009 TED talk on the effects of hardcore pornography as default sex education—and her answer, the social sex videosharing platform MakeLoveNotPorn.[4]

---

4   Cindy Gallop, "Make Love, Not Porn," TED Blog, December 2009.

Even before there was a name for the category, entrepreneurs like Wendy Strgar, Rachel Braun Scherl, Karen Long, Amy Buckalter, and Ti Chang were building companies and developing products like Good Clean Love (the first organic personal lubricant), Semprae Laboratories's Zestra (a topical product to increase arousal), Nuelle's Fiera (FDA-cleared personal care device to help increase arousal), Pulse (personal lubricants delivered by a warming device using a patented induction heating process and clean delivery system), and Crave (tech-savvy sex products like vibrators), respectively.

Of course, what makes headlines today are more often sensational topics—futuristic lifelike sex robots, virtual reality porn, crypto sex worker payments. Journalists assume that if it's about sex, it has to be sexy.

While the idea of a fully realized AI lover makes for a compelling read, much of what has been written about sextech is incredibly narrow. Many sextech entrepreneurs—myself included—see a much larger opportunity in addressing sexual health and pleasure, closing the orgasm gap between men and women, and addressing the needs of populations whose needs have never been the subject of serious innovation.

But this can be hard to do when we barely teach that the

clitoris exists. We've come a long way, but there's still so far to go.

SEXTECH VS. SEXUAL WELLNESS

While the "sextech" label is generally applied to technology and innovation, you'll often hear it paired with another term—"sexual wellness." Sexual wellness encompasses all companies serving people's sexual needs, including medical, pharmaceutical, healthcare, mental health, media, and other innovations, technologically based or not. Sexual wellness and sextech generally refer to products and services that focus on pleasure rather than just reproduction.

It's hard to overestimate the market for sextech—or the difficulties in getting past the stigmas that come with it. As I'll discuss later, sexual wellness entrepreneurs face an uphill battle in bringing products to market.

The sexual wellness market is fragmented and—shockingly for retail—still significantly offline. There aren't any public companies in sextech, and I have spent years tracking down reliable sources of information, combing through industry reports, and speaking to executives and other insiders to get a better picture. Industry insiders that I have spoken to have estimated that Adam and Eve and LoveHoney, two of the biggest e-commerce brands, do about $100M-$150M each in annual online sales. Amazon, who notoriously keeps their data under wraps, did a reported $339M in online sales in 2014.[5] It is the biggest e-commerce provider of sexual wellness products, with approximately 60,000 products. And yet, today, I'd estimate that they still likely sell just over $1 billion in sexual wellness products, about 1/20th of total global sales.

There are very few dominant brands or established marketplaces, and Amazon isn't poised to win the space.

But make no mistake—the market is there. In 2019, the market for sex toys alone is valued at $27 billion dollars globally. If you consider all sexual wellness products, including lubricant, erotica, and out-of-pocket contraceptive costs, globally sexual wellness is valued around double that at $50 billion. Some experts predict the sexual wellness market to grow as high as $122B by 2026, with a 13 percent CAGR (calculated annual growth rate).[6]

My investors may not know the structure of the clitoris, but they sure understand those numbers. And nearly every investor I speak to is intrigued by the space. Many

---

5   Kline, "Sexual Wellness: U.S. Market Analysis and Opportunities," 2014.

6   Reuters, "Sexual Wellness Market 2019 Global Analysis, Opportunities And Forecast To 2026," February 2019. Estrella Jaramillo, "Investing in Sextech: Two Founders Breaking Barriers Internationally," June 2019.

are wary of missing out on what may be one of the last great untapped retail markets in the world.

Honestly, they're nervous. Investors are largely older men, which means they sometimes don't understand the problems to be solved. Or they answer to more conservative institutions that won't invest in "vice." Stigma keeps them from sextech.

The struggle to recognize the importance of sexual wellness is in some ways akin to the struggle that mental health faces. As with mental health, sexual wellness asserts that our relationship to sexual health has direct connections to our overall health, and needs to be taken just as seriously.

THE EMERGING FUTURE

Unfortunately, nearly every aspect of society today—public health organizations, school-based sex education, financial institutions, investment and venture capital, major social media companies—still rejects this very basic reality. At best, it's considered indulgence. At worst, sexual wellness and technology are routinely grouped along with pornography, illegal drugs, weapons, and gambling. Society has deemed what sexual wellness companies do as "immoral" and therefore demands it to operate underground and in the dark.

For entrepreneurs, innovators, and investors in the space, this makes the work especially tough. Those of us who work in the space do it because we're passionate, we're driven—and we know that if the gamble pays off, the rewards, financially and societally, are immeasurable.

As a millennial entrepreneur, I realize how quickly the world is changing. Another sexual revolution is on the horizon—one that's more inclusive, more diverse, and more challenging for older companies to navigate.

Ten years ago, investors in the private sector were barely willing to fund menstruation care, breast pumps, or fertility apps, products in a space referred to as "femtech." In 2012, femtech companies had only raised $62 million. However, companies in the femtech space have broken records, with a projected funding of $1 billion in 2019.[7]

Sextech is not far behind. While stigma about pleasure and sexuality still forms a formidable barrier, experts and researchers point to an opening mindset and shifting sexual needs. For the first time in American history, single women outnumber married women.[8] Increased

---

7   Dana Olsen, "This year is setting records for femtech funding," PitchBook, October 31, 2018.

8   Rebecca Traister on Fresh Air, "Single By Choice: Why Fewer American Women Are Married Than Ever Before," NPR, March 2016. Rebecca Traister, "Single Women Are Now the Most Potent Political Force in America," The Cut, February 2016. Shamard Charles, MD, "Teens Are Having Sex Later, Using Contraception, CDC Finds," NBC News, June 2017. Joyce Abma and Gladys Martinez, "Sexual Activity and Contraceptive Use Among Teenagers in the United States, 2011-2015," CDC, June 2017.

access to pornography means that subjects that were once taboo are now openly discussed. From Fifty Shades of Grey to Broad City to Grace and Frankie, we're slowly and steadily getting more accustomed to acknowledging—if not discussing—sexual pleasure. People who stay single longer have more sexual partners and buy more sex toys—and studies show that younger women are trying sex toys in droves.[9]

From gay marriage to trans awareness, society is steadily becoming more aware that there are mental, emotional, and physical downsides to denying, hiding, or silencing our sexuality. To me, this signals a major turning point. Since the Sexual Revolution of the late 1960s, society has been willing to talk about sex for titillation or laughs— finally, we're talking about it in terms of health.

Think about this: Millennials are twice as likely to identify as LGBTQ as previous generations, and one-third of Gen-Z people know someone who uses gender-neutral pronouns like they/them. For these newer generations, shame-based views about sex and sexuality are increasingly outdated.[10]

---

9   Bella DePaulo, "There's never been a better time to be single," CNN and The Cut, March 2018.

10   GLAAD, "New GLAAD study reveals twenty percent of millennials identify as LGBTQ," March 2017. Kim Parker, Nikki Graf, and Ruth Igielnik, "Generation Z Looks a Lot Like Millennials on Key Social and Political Issues," PEW Research, January 2019. NPR, "The New Sexual Revolution: Polyamory On The Rise," February 2019.

Even Boomers are embracing their sexual lives. Online dating and advances in hormonal and erectile dysfunction therapies have helped put to rest ideas that seniors are sexless. Even though the average frequency of sex decreases as people age, menopausal aged women are reporting higher levels of sexual satisfaction.[11] We're seeing more attention around menopause and painful sex, which means more attention on lube and other sexual wellness products. Every generation with purchasing power is getting more and more ready for sexual wellness and sextech industries to meet our changing needs.

At the same time as we're expanding our notion of who can be sexual, we're increasingly left without the resources to deal with it. Public health agencies struggle to address sex outside of the context of STIs or pregnancy, sex education in schools is at an all-time low, and online access to porn—often to kids as young as nine years old—is at an all-time high. Conversations around sexuality, gender, reproductive rights, and consent often reveal that we lack a common language for communication.

For those of us who work in sexual wellness, who see the potential for sextech, this is our moment.

We have the power to change this. To create a world,

---

11   Dr. Pepper Schwartz, "Baby Boomers Getting Older But Not Giving Up on Sex," American Sexual Health Association, Accessed June 2019.

empowered by tech, public health advocates, and innovation—that serves the needs of all genders, all sexualities, and all ages. To integrate sexual wellness into our medical system. To be able to talk about sex and sexual pleasure without embarrassment, euphemisms, or guilt. We owe this to the next generation to fix what was broken in ours—their safety and well-being are more important than our shame.

I know, because when it came to sex, fear and shame were all I was taught.

HOW DID I GET HERE?

My mother was seven months pregnant with me when she immigrated to the United States from the Philippines to join my father in Sacramento, California. Soon after, my little brother and sister were born. We were raised in a very loving, but strict, traditional church-every-Sunday Filipino Catholic home. The only thing I was ever taught about sex was not to have it—at all—until marriage.

I was taught that men would try to get sex from me, and that my job was to refuse them at all costs. During Sunday school, I learned the sacred importance of virginity, and I was taught that homosexuality was a sin against God. In public school, I was shown graphic photos of STIs, a long video of a live birth called *The Miracle of Life*, and also simulated social experiments to teach us the dangers of having multiple sexual partners.

As a result, I spent most of my adolescent years and early twenties in the dark about sexuality, in denial about my early attraction to women, ashamed of my desires, completely shut off from my body, and, consequently, shut off from much of my power. I coped by focusing all of my repressed energy into school and work.

At the age of sixteen, I enrolled at Bard College at Simon's Rock, and after transferring, I graduated from UC Berkeley with a degree in linguistics and a plan to move to China. Then, my freshman roommate, Jessica Mah, called me out of the blue. She had just raised over one million dollars of investment from YCombinator and other well-known Silicon Valley investors. She was calling to ask me to be her right hand, and help her build an accounting software company called inDinero.

I was incredulous; at twenty years old, I knew absolutely nothing about business, accounting, or technology. But something about Jess' guts—and her huge mission— inspired me. I dropped everything and joined her in Mountain View, California, in the late summer of 2010. From accounting operations to sales psychology to management, I spent the next three years learning every day and helping hundreds of other companies manage their finances and accounting. Today, inDinero is one of the leading financial solutions for growing companies. I am where I am today because another woman dreamed bigger for me than I dreamed for myself.

In 2014, I was tapped to join a venture capital seed fund named 500 Startups as a venture partner. At inDinero, I operated a single company with one specific business model. 500 Startups exposed me to hundreds of technology startups, business models, and gave me massive insight into the inner workings of the venture capital ecosystem and culture. In addition to expanding my worldview about technology's impact on the world, positive and negative, I helped deploy millions in venture capital to dozens of startups, heard over a thousand tech startup pitches, coached founders to raise many rounds of venture funding, and I invested all over the world, including West Africa, Brazil, Southeast Asia, and the Silicon Valley.

On the outside, it looked like I had everything going for me, but privately, my repressed sexuality started to take its toll. I was struggling with my identity and mental health. No matter where I looked, it seemed like there was a huge gap in resources. It was so ironic that I spent most of my days talking about the internet, but the problems I was struggling with the most weren't being solved online. My experience learning about sexuality on the internet was full of harassment, toxic comments, and being sent unsolicited pictures of penises.

The first time I had a positive, empowering experience was the first time I visited the Good Vibrations in Berkeley, California, a popular Bay Area retail chain that sells sex toys and other sexual wellness items in a clean, open, unashamed space, with knowledgeable salespeople and an on-staff licensed sexologist. Joani Blank's vision of what a sexuality retail experience could feel like would inspire a movement of education-first sexual wellness retail stores all over the country, as well as my own vision for the future of sexual wellness.

IT ALL STARTED WITH GOOD VIBES

If you want to know where sextech started, it's worth a visit to the Antique Vibrator Museum in San Francisco. Managed by Good Vibes staff sexologist Carol Queen—the Museum is actually attached to a Good Vibes location—the Museum teaches us a lot about how the sextech industry came to be in the years before Joani Blank.

Clitoral stimulation tools, like vibrators, have a surprisingly long history. The first were actually invented in the 19th century—before even the vacuum cleaner. In fact, the first vibrators were steam-powered. They were not marketed as masturbatory aids, but medical ones. They were meant to relieve "hysteria" in women. (Previously, doctors had done the job with their hands.)

George Taylor, an American physician, patented the first of such devices in 1869, but it would be decades before their usage would extend to the general public, once electricity became a common fixture in middle-class homes. Battery-powered vibrators and electricity helped fuel the popularity by 1900. Perhaps not coincidentally, women were the biggest consumers of electrical appliances in the home.

Over the next few decades, ads that could today be described only as tongue-in-cheek bombarded publications that targeted homemakers. According to these ads, vibrators were "invented by a woman who knows a woman's needs," and "...furnish every woman with the essence of perpetual youth."[12]

Sadly, once vibrators became seen as the domain of sexual pleasure—particularly the use of vibrators in erotic media—mainstream publication would no longer carry vibrator ads. Ads for "massagers" returned in the 1950s, but weren't openly marketed as sex aids until the 1970s, when adult media like *Penthouse* and *Oui* developed to take them, and adult bookstores to sell them. By 1976,

---

12 GoodVibrations, "Antique Vibrator Museum." Accessed online, August 2019. http://www.antiquevibratormuseum.com.

pornography distributer Reuben Sturman partnered with traveling salesman Ron Braverman to establish Doc Johnson—a sex toy company that would become one of the largest in the world. But the intended market for those toys was often men, purchasing them on behalf of women, and the quality left much to be desired.

It was Joani Blank who opened the first Good Vibrations store in San Francisco, serving as a safe space and bastion of sex-positivity for the modern woman—and an inspiration for me and many others in sexual wellness.

When I visited Good Vibes for the first time in 2008, I was struggling with my sexual identity and eighteen years of accumulated shame. The first thing I noticed was the store lighting. Unlike other adult stores, the space was well-lit. Vibrators, lube, condoms, and other items were showcased unapologetically, curated with care and pride. The store also had a dramatically different vibe—the other patrons were overwhelmingly women and queer people and the staff was helpful with a positive demeanor. I was too shy to ask any questions directly, but I overheard cheerful, highly trained explanations of how one could safely choose a butt plug and patient demonstrations of clitoral vibrators—all from people who seemed to emanate safety and acceptance. It was a healing experience, and it also showed me what could be possible for me, not to mention the sexual wellness space.

That first purchase helped me learn more about my body, and more about who I was as a person. I understood how integral my sexual identity is to my overall identity, and realized the toll that denying it or hiding it takes. Over the years, this nagged at me. Here I was, a millennial in progressive San Francisco, lucky enough to have a resource to help me explore safely and without shame. Surely people all over the country, and all over the world, facing even greater hurdles, were suffering as well. What would have happened to me if I had been born in a state without a Good Vibrations or another safe space?

On July 25th, 2016, a few days before I resigned from my venture partner role at 500 Startups to build something in the sextech world, I wrote Joani an email:

> Dear Joani, I am a huge fan of yours. Thank you for bringing Good Vibes into this world. It's truly an inspiration for me.
>
> I'm writing because I am starting a technology company in the sexuality space, and I would love to meet you for coffee. Are you free next week, maybe the week of 8/8?

Joani Blank passed away from pancreatic cancer on August 6th, 2016, just a week after I reached out for coffee.

I doubt she was able to read my email, but if I could have met her, I would have thanked her for paving the way and designing a space so transformative for me and many others. Humbled and inspired by Joani's passing, I started my own journey and set out to build a sextech startup that could reach billions.

THE SPACE BETWEEN PLANNED PARENTHOOD AND PORNHUB

As I prepared to launch my next company, I became consumed with curiosity and determination, and saw a market opportunity that no one around me seemed to see. I became a power consumer of everything sexuality related, online and offline. I discovered a content company called OMGYES which produced high quality, nude demonstrations of clitoral stimulation videos; a controversial organization called OneTaste in San Francisco that taught "orgasmic meditation" courses; an Ohio-based "Tupperware party"-esque multi-level marketing company called Pure Romance employing tens of thousands of party consultants; and a huge swath of online websites and resources, spanning from respected sites like Scarleteen to a plethora of sex blogs and podcasts.

There was little online that wasn't overly medical or outright misinformation—very little between Planned Parenthood and PornHub. There needed to be a space online where people could learn about sexuality from curated, high-quality experts who would not judge or shame anyone and who would be inclusive of people of all sexual identities, genders, and backgrounds.

The first question that I had was, how?

There were at least four million sex-ed YouTube videos when I started researching the problem. I discovered that curation was a big problem, because what the algorithm recommended wasn't necessarily what you needed. There was very little quality control and, besides, recorded video didn't feel as transformational. I wanted to mimic what worked in offline settings: participating, even silently, in groups of people who shared your same problem. These were the types of places I felt my own shame melt away, and I was bullish that scaling the feeling of a judgment-free space was a big opportunity.

At 500 Startups, I studied business models like Udemy, Coursera, Netflix, Etsy, and tried to think of the audience I wanted to serve. Twitch—an online streaming and chat platform that built its success on gamers—was also a model I knew very well coming from a family of gamers. What was the best way to disseminate accurate information about sex, pleasure, and sexuality, and create transformative experiences? That's when I decided to start seriously exploring live streaming as a possible solution.

My first step was reaching out to sex educators. I met

with the most dynamic sexual health educators, spanning many topics and community problems. I met an amazing educator named Latishia James, "Reverend Pleasure" on O.school, who had just completed her Masters in Divinity from UC Berkeley. She taught me about sex education at the intersections of race, religion, and reproductive justice, especially through the lens of sexual and religious trauma. I learned about consent from Eva Sweeney, a disability and sexuality expert who was non-verbal due to cerebral palsy. From certified sex and dating coaches I learned about navigating modern dating apps, former professional dominatrices taught me about relationship boundaries, Good Vibrations educators showed me the nuances of sex toy safety education, and ex-Planned Parenthood educators detailed for me just about every flavor of sex education. I learned from kink professionals and adult performers, and immersed myself in the world of sexuality.

It was a foreign world to me, and after a few conferences, I was vexed to find that much of the education happened offline at spaces, usually without any online components. I was also disappointed that none of the sexual influencer brands seemed to break through to the mainstream. I wondered if creating a space for this live education was how I could contribute.

In order to recruit a team and start building the live

experiential model, I needed to raise some money. In Silicon Valley, the first round of funding is often called the "friends and family round." Coming from a lower middle-class immigrant family, I didn't know anyone whom I could ask for money. Actually, my family at the time was pretty perplexed at my decision to leave my stable venture capital position. So, I started taking my friends to coffee. All I had was a pitch deck and an idea, and many people told me it would be near impossible to raise traditional funding in this sector. However, after about eight weeks, I was able to raise a few hundred thousand dollars from individuals like Lars Rasmussen, the co-founder and inventor of Google Maps; Cyan Banister, one of the best angel investors in the Silicon Valley; and my previous employer, 500 Startups. That was the first step—proving there were great investors who would back this.

Next, I built a prototype with a designer and an outsourced engineering team in Serbia. We signed up our first group of educators and began running experiments. During this stage, a few of our investors and advisors took notice. One of them joined the company as one of our engineers. Another joined as my co-founder, Kelly Ireland, a serial founder and engineering leader at WebMD. This was another important step—proving there were talented people who would want to work on this project with me.

Our initial livestreams were everything I had hoped—

engaging, informative, and positive. We were soon attracting users from all over the world. Some spent up to three hours a day interacting with us through live chat. The demand was there, but the model we built, a livestreaming network with live sex educators streaming over forty hours per week was a huge lift for the team. The proof of concept was there—we were helping people—but it was hard to keep it constantly staffed, and to make it work financially for the educators, who were dependent largely on tips. The biggest problem we also encountered was the difficulty growing an organic audience. It was my introduction to the biggest challenge of any sextech venture: customer acquisition. We went back to the drawing board.

My first year of building O.school, I was obsessed with how to help people transform their lives. Live streaming proved to be highly engaging, but failed to organically attract new users. So while we got press attention, that word-of-mouth customer acquisition did not grow the way we needed it to. It was clear that we needed to build content and community that existed anytime you needed it—not just when someone was streaming.

But we noticed something happening organically—our sexual health content was beginning to rank on Google. The search engine had determined that, at least on some topics, we were reputable—or even the most reputable.

We began to shift directions to focus on building medically accurate sexual wellness content—to be the best and most trusted sexuality brand in the world. We have a long journey ahead to reach 8 billion people, but we have a path. And a couple more million dollars in investment funding to build a brand in the wide open space between Planned Parenthood and Pornhub.

And we are just getting started changing the way all people, especially young people, search for and experience sexual wellness content online.

{ CHAPTER 2 }

A CLOSER LOOK AT SEXUAL WELLNESS

In the past five years, I've spoken to thousands of students at colleges and universities across the country, from massive auditoriums at state schools to small, intimate gatherings at liberal arts colleges. I've spoken at tech conferences, like TED Unplugged in Vancouver, SXSW in Austin, and the UN Commission on the Status of Women in New York City. And, of course, I've hosted livestreamed classes on O.school. No matter what the venue, the reaction I get when we talk about pleasure and sex education is incredible. When we're able to solve a problem, the results can be life-changing.

But it's not just in formal settings. Whenever I talk about what I do, people confide in me.

I'll never forget a fifty-six-year-old woman at a business networking event telling me that she finally had her very first orgasm—after learning how to use a Hitachi Magic Wand, a mass-market massager beloved by multiple generations.

Or the Lyft driver in 2016 who burst into tears trying to communicate to me the shame she felt around sex. She'd had four children, she told me, but her husband didn't care about her pleasure. Desperate to have someone to talk to, she finally asked me through tears if it was okay to buy her fifteen-year-old a vibrator—she wanted a different life for her daughters.

Or the law school student at an Ivy League university who came up after a talk who told me she had no idea that women could really experience sexual pleasure. She'd seen it on TV, of course, but assumed it was a myth because it had never happened to her.

Or the disabled veteran in Kansas City who cried to me about his inability to please his wife due to post-traumatic stress disorder that affected his marriage and relationship to intimacy.

Or the sorority member in Los Angeles who thought she was broken. She'd had sex with many men but never had an orgasm.

Or the Muslim trans man seeking help about coming out to his family. His older sister had been kicked out of their house for coming out as a lesbian, and he was desperate for insight as to how I managed to reconcile with my own religious parents.

We tend to internalize our sexual shame. I've had women in their fifties—CEOs and leaders in their fields—approach me after a business meeting to confess that they fear they're failing their husbands, because they feel disconnected from their bodies.

The unifying thread in all of these stories is shame. Shame can have many causes, but I frequently find religious upbringing, sexual trauma, miseducation, and lack of education to be the key drivers. People are very often ashamed about their lack of knowledge, despite the fact that we as a society have failed to educate them. In place of education, we've created an ideal of a sexually-realized woman from *Sex and the City* to *Broad City* to *Cosmopolitan* to porn, we've created an ideal—progressive in some ways—that can paradoxically leave women feeling like sexual failures.

So when I talk about sexual wellness to strangers, when I talk about the ways in which we've failed to educate, and how the path to pleasure is different for each person, it can be hugely validating and start a process of healing for these deep wounds.

People think sexual wellness is about sex, when it's really about wellness.

We're going to talk a lot about sexual wellness as we get into sextech, so here is what I believe are the core principles:

Sex is natural and normal. Sex, masturbation, and sexual pleasure are regular functions of the human body. Pleasure is not dirty or shameful, indulgent or harmful, nor to be shrouded in mystery.

It is okay not to have sex, too. For some people, sexual wellness means not being sexual. Celibacy and abstinence can be valid, healthy choices, especially for individuals who identify as asexual, about one percent of the human population.[13]

Sexual wellness is part of health and wellness. As such, it should be in the domain of evidence-based public health—not religious, political, or morality-based institutions.

Sexual wellness is not "sexy." We should be able to address it without sensationalism or "hot" models. You don't need to act sexy to educate people about sex.

---

13   AF Bogaert, "Asexuality: prevalence and associated factors in a national probability sample." *J Sex Res.* 2004 Aug;41(3):279-87.

Sexual wellness is holistic. It uses mental health, spiritual health, physical health, and sexual health to connect people to their own sexuality.

Sexual wellness is not about performance. There's not one way to be a sexual being. We don't have to meet others' expectations about what a fulfilled sex life looks like.

Sexual wellness products are not "novelty" or "niche" products. Vibrators, cock rings, lubricant, and other products should be treated the same way as a leg brace or a back massage.

Sexual wellness is about access. We deserve education and the freedom to make informed choices about the sex life that is right for each of us.

SEXUAL PLEASURE IS GOOD HEALTH

As we will explore in the next chapter, most of us are taught about sex in negative terms—dangerous STIs and unwanted pregnancies, ruined reputations, and legal liabilities. As kids, we're taught the "why not" of sex, but rarely the "why." Not only is that confusing, but for many of us, it has lasting impact on how we approach sex as adults. After all, most of the sex that people have is not for reproduction, but because it feels good. Because we're attracted to someone, or because we want a more personal connection, or because our body is craving sexual release. But it has a host of other benefits as well.

For most people, sex reduces stress. In addition to just feeling good, pleasurable, consensual sexual activity releases oxytocin, a hormone which creates feelings of well-being, facilitates social bonding, and helps block certain stress reactions in the body.[14]

Sexual pleasure also releases endorphins—the same hormones that kick-in with exercise. Sex helps with sleep, which can in turn affect everything from mental health to longevity.[15] It has even been shown to strengthen immunity.[16]

We face an epidemic of stress-related illnesses, from anxiety to heart disease, and have become used to doctors scolding us about what we eat and if we work out—but when was the last time a doctor asked about your sex life, or if you masturbated regularly? In most cases, such a discussion about sex—one of the key drivers of human behavior—would be entirely off the table.

---

14 Kerstin Uvnäs-Moberg et al, "Self-soothing behaviors with particular reference to oxytocin release induced by non-noxious sensory stimulation," *Front Psychol.* 2014; 5: 1529.

15 Michelle Lastella et al, "Sex and Sleep: Perceptions of Sex as a Sleep Promoting Behavior in the General Adult Population," *Frontiers in Public Health,* 2019; 7: 33.

16 Carl Charnetski, Francis Brennan, "Sexual Frequency and Salivary Immunoglobulin A (IgA)," *Psychological Reports,* June 2004.

Sexual wellness challenges that model. Sexual wellness seeks to strip away the prescribed versions of sex that society has given us—what we should or shouldn't do, who we should or shouldn't be—and to integrate it back into our everyday lives.

Sexual wellness companies understand sex is a source of energy, power, and self-care. A sexual wellness company incorporates sexual health, mental health, and general wellness in order to help someone feel at home in their body, free of psychological and physiological issues around sexuality.

We celebrate finding out the diet that works for you and doing the exercise that's right for you. We remind each other to get our water intake in order and to sleep eight hours. But how often has the medical establishment told us to get your sex life and sexual identity in order?

That's what sexual wellness calls for. It's bringing sexuality back into the experience of being human.

ERECTILE VISIBILITY, VAGINAL SHAME

Turn on the Superbowl, and you'll see ads for Viagra, Cialis, and Levitra, all drugs treating erectile dysfunction (ED). Ever since former Senator Bob Dole appeared in a Viagra ad, it's been acceptable to talk about ED on television.

In general, this is a good thing. All sexual problems are valid and should be respected, and I've seen firsthand how people struggle with ED and the emotional toll it can take. We want more discussion about sexual problems, not less.

However, the near-singular focus on ED is representative of how the medical industry addresses sex: in terms of performance, rather than pleasure or satisfaction.

With the generic form of Viagra now available, ED advertising has exploded seemingly overnight. And with it, the idea that erectile performance is a performance sport.[17] Erectile dysfunction is a serious condition, but the barrage of ads—on radio, on bus stops, on social media—not only focus entirely on one gender, and one form of sex, they focus only on one solution.

We've all heard it in ads: "Call your doctor for an erection lasting more than four hours." In a culture that pushes men to last longer and perform better, even the warning of "side effects" is a form of marketing.

But many doctors and sexual health experts recognize that ED is often more complicated than a lack of blood flow to the penis.

---

17   Ian Crouch, "Viagra Returns to the Bob Dole Approach," *The New Yorker*, October 2014.

Stress and anxiety, and other mental health concerns, are a big factor not only for ED, but premature ejaculation, the other even more common penis-related sexual concern. This means that the pressure that these ad campaigns promote—that good sex must include an erect penis that lasts for a long time—can actually exacerbate the problem.

Men under thirty are increasingly reporting erectile dysfunction, but some experts believe that this is less an increase in incidence, as much as an increase in expectation that erections must be unwavering.[18] Part of the issue is the failure of our sex education system, which we'll talk about more in chapter 3, and expectations set by porn, which we explore in chapter 4.

Where the medical establishment seeks to pathologize and prescribe, sexual wellness recognizes that sexual performance is a complicated issue that involves not only physical reactions but mental and emotional ones.

But selling that business model to an investor is a bit more difficult.

It should come as no surprise that funding for pharmaceutical solutions to sexual issues have outpaced every

---

18   Rastrelli Giulia and Mario, Maggi, "Erectile dysfunction in fit and healthy young men: psychological or pathological?" *Transl Androl Urol.* 2017 Feb; 6(1): 79-90.

other segment of sexual wellness, and sextech more specifically.

As of 2019, the only unicorn—a company with an estimated market value of over $1B—in sexual wellness is Hims, a company that is known for providing generic ED prescription via chat-based consultation, and its companion brand Hers, which provides prescriptions for Addyi, the recently approved but somewhat controversial "female Viagra."

A business model based on the easy sale of drugs like Viagra and Addyi is not the answer to sexual wellness, unfortunately. They may be part of the solution, but without more holistic attention to the problem, they may pose public health risks.[19]

But here's where sextech entrepreneurs can take heart: the erectile dysfunction market may be easier to access, but it's still small in actual size. ED is projected to be just under $3 billion in the next decade, according to some estimates, while sexual wellness, including vibrators and other pleasure-related products is projected at a staggering $27 billion.[20]

---

19 I explored this topic further in, "Warning: The Sex Drug Startup Revolution May Have Side Effects," *Forbes*, December 2018.

20 Jackie Rotman, "Vaginas Deserve Giant Ads, Too," *New York Times*, June 2019.

Ironically, companies like Hims may be quicker to erect, but they may not have the size or staying power of a more holistic approach. And as we'll see in future chapters, the playing field is much more open.

BEYOND NOVELTY ITEMS

So why is an industry valued at $27 billion globally being ignored in favor of one nearly ten times smaller? Because we don't take sexual wellness seriously.

Historically, vibrators, dildos, and other sexual devices were referred to in the retail trade as "novelty items." They were sold in the back of "dirty" magazines, or in adult bookstores, near crotchless panties and French maid outfits. In some states, the sale was illegal. In others, it was merely hidden. How could anyone take them seriously?

Yet ask anyone who has relied on a vibrator in order to climax during sex, or as a trusted masturbatory aid, and they will tell you sexual wellness is far from a "novelty" in their lives.

People who have struggled with pleasure and orgasm have shared with me how their a-ha moments around orgasms, masturbation, or a new fulfilling sex life helped them unlock confidence in their careers, overcome traumatic experiences, improve virtually every kind of relationship, save marriages, and build a better quality of life.

Andy Duran, education director of Good Vibrations, often shares stories about customers who, at first, come in the store, quiet and dressed to hide their identities, only to return with completely changed demeanors and effusive gratitude to the employees who helped them with their purchases. He told me, "I've been helping people find pleasure through sex toys for almost half of my life. Having conversations with complete strangers seeking support in one of the most intimate aspects of their lives has honestly been one of the most rewarding experiences of mine." How many other types of purchases can cause customers to return to thank the employees who helped them?

Not everyone has an interest in or need for a sexual wellness product like a vibrator or dildo, but as they've become less stigmatized and as women's sexual pleasure is more recognized, more people are finding them to be a key to sexual pleasure. From Fleshlights to Magic Wands, I have met many people who have held ceremonies to retire their trusted companions.

As our bodies change, so can our need for products like vibrators and lubricants. As Amy Buckalter, the founder of lubricant company Pulse, says, the latter can become a hugely important part of sexual life when "the faucets dry up."

Amy bemoans the use of the term "sex toy." She says it reinforces the idea that sexual wellness products are frivolous gag gifts found next to the fake puke and sexist T-shirts in a strip mall Spencer's.

The idea behind Pulse was born while Amy was herself approaching menopause but was dissatisfied by the quality of products out there. Pulse has developed its own silicone and water-based lubricants paired with a warming device "that uses a patented induction heating process and clean delivery system."

In an Italian restaurant, Amy pulled out the warming device to demo the products for me, and women on both sides of us stopped their dinner to talk to us. I watched Amy beam with pride when the women praised the sleek design of the warming device. Moments before, she explained to me that she wanted to design something as elegant as an Apple product, which she felt was missing from sexual wellness products.

Everyone has to learn what feels good for their body. Unlike what we're told by Hollywood, an orgasm does not have to be the goal during sex, but everyone who is sexually active should be given access and support needed to orgasm. Upwards of 70 percent of clitoris owners need or prefer clitoral stimulation to experience orgasm[21]—the dominant depiction of intercourse, penetrative penis-in-vagina sex, will not result in a climax without additional stimulation—but you wouldn't know it from most mainstream media.

In *Come as You Are*, Emily Nagoski describes libido not as a "sex drive," but rather as a gas pedal and brake. There are forces, internal and external, in our control and out of it, that affect how we perceive sex, and our desire for it. It changes according to age and partner and time of life. It is not linear.

SEXUAL WELLNESS IS NOT "SEXY"

"They aren't hot enough," an investor tells me. Again.

"Who?" I ask, knowing I'll regret to hear the answer.

He could be talking about the educators at O.school (all of whom were beautiful, but were not white, young, or thin) or those we've included in an educational campaign. He expects supermodels.

---

21   Debby Herbenick, "Women's Experiences With Genital Touching, Sexual Pleasure, and Orgasm: Results From a U.S. Probability Sample of Women Ages 18 to 94," *Journal of Sex & Marital Therapy*, August 2017.

"But women don't always want to learn about having their first orgasm, body shame, or having sex after giving birth from supermodels," I tell him. He listens but doesn't understand. But sex sells, you can see him thinking.

I'll get into the gender gap in venture capital later in the book, but the desire for "hotter models" isn't new. After all, that's how it's done everywhere from Carl's Jr. to Hustler Hollywood. The traditional "novelty" business still sells sex toys with porn star packaging.

This may have been an effective strategy to attract male consumers—those who traditionally made the purchases in adult bookstores and through mail-order—but sexual wellness is more and more driven by women consumers.

Except...sex doesn't sell sexual wellness. In fact, it limits it. If you look at the leaders in the high-end sexual wellness product space—companies like Maude, Dame, even Hers—the marketing isn't particularly sexy. It doesn't shy away from sex, but it's not...breathy. There are no fake orgasms or arched backs, just innovative, sleek solutions to a wide variety of needs.

As Rachel Braun Scherl—a self-described Vagipreneur, the author of *Orgasmic Leadership* and innovator behind Semprae Laboratories' Zestra topical product for arousal— joked to us in a conference audience, "Ask anyone who has had 'vaginal dryness' and they'll tell you it was the least sexy thing imaginable."

While it can be tough to communicate to an investor accustomed to Victoria's Secret, sexual wellness is different. Women don't need, or even necessarily respond well to, nudity or sexy images when they are buying sexual wellness products. Women are too often desexualized or hypersexualized according to age, body size, sexuality, gender presentation, or race. Using a thin, light skinned, twenty-three-year-old woman only tells a sexual wellness consumer that this isn't for her.

Presenting a range of beauty, bodies, and desirability to a consumer isn't just ideologically sound—it's better business.

FIRST YOGA, THEN MEDITATION, NOW...

One day soon, buying a vibrator will be viewed like buying a yoga mat.

Ten years ago, few people would have believed you if you predicted there would be massive, venture-funded companies dedicated to meditation and mental health. Meditation was seen as a niche space. Mental health was something dealt with by prescription, not a computer. And yet the success of companies like Headspace, Calm, and Talkspace have all stunned investors.

In the past twenty years, wellness has moved into the mainstream. Once fringe ideas—yoga, pilates, juice bars—are now huge businesses.

Our puritanical moral standards make it hard for many to see how sex can move mainstream, but wellness offers a lesson. By shifting the message away from prurience and toward wellness, it allows us to access a much broader market.

Think about CBD oil, the non-psychoactive (and legal) cannabis derivative that is said to treat everything from anxiety to pain. Five years ago, only dedicated potheads knew the letters. Marijuana wasn't short on stigma—after all, it's still a Schedule 1 drug, and possession carries a one-year prison sentence. But medical hemp and CBD products are expected to reach $27B by 2026.

That's the power of wellness.

As we'll get into later, the generation that grew up with meditation apps and a focus on self care has also learned how to bring sexuality into their overall health. The more awareness and acceptance people develop, the more empowered they feel to take care of their sexual wellness as well.

We're not new, but we're the next frontier.

ENDING SHAME

Shame is the single most powerful force holding back sexual wellness. Shame spirals into fear, which in turn keeps people from talking about these subjects altogether. Fear, in turn, begets stigma—and keeps people from wanting to participate, fund, support, and partner in these spaces.

But stigma isn't invincible, and neither is fear or shame.

We can overcome the counter forces of shame, fear, and miseducation. We can find ways to deny conservative forces the ability to dominate public policy. We can fight for evidence-based science, sexual health experts, and public health professionals.

By proving the market, sextech entrepreneurs are beating a path to success. We're showing that it's not just sustainable, but hugely profitable. And if there's one truth in America, it's that public morals are no match for private equity. I'm not saying market forces are always good—far from it—but from Prohibition to PornHub, the easiest way around conservative forces has always been a clear profit.

In order to get there, we need to know what we're up against. In the next chapter, I'll explain what's keeping us in shame and what we need to do to fix it.

{ CHAPTER 3 }

THE SEXUAL MISEDUCATION OF AMERICA

Every time I give a sex education talk or workshop at a university, I put out a box where the audience can scribble down questions anonymously and submit them for me to answer. The questions are incredibly telling, often more so than the questions asked in public.

The questions people ask out loud are often advanced, and sometimes performative—perhaps dealing with polyamory or the ethics of porn. If you were to sit in on one of these, you'd be impressed with the nuance many young people bring to discussions about sex. But the questions in that answer box are different.

Some questions I've gotten recently:

- Will I go to hell if I am not a virgin?
- Is sex supposed to hurt?
- How do I masturbate?
- Do I have to shave my pubic hair?
- How does gay sex work?

These are incredibly basic questions, but the students aren't stupid. In fact, some of the questions above have been submitted to me by students at Ivy League universities.

As sextech entrepreneurs, we tend to concentrate on the people who ask the questions out loud. The bold, the progressive, the kinky. But the vast majority of our audience, the ones who really need us, are the ones who submit the questions in the box.

The issue isn't intelligence—it's that we don't teach about sex.

At smaller state schools, the questions I get asked are a direct window into the quality of sex ed that a region provides. Once, after a talk at a small school in Kansas, a young Indian student stood and explained that a nurse had told him not to masturbate—that it could lead to erectile dysfunction.

The educator I was with explained to him how wrong that

nurse had been—that masturbation is a healthy practice, that pleasure is a natural physical reaction to stimulation, that the nurse's advice wasn't rooted in medical science at all, but instead moralistic stigma.

He sprinted out of the auditorium.

His story isn't unusual. Across the country, students are routinely taught regressive, harmful, medically incorrect information in schools. Most states don't require that schools provide any sex education to students, meaning it's left up to the school districts.

As I've discovered first hand, it's a major crisis.

We now have a system in which your knowledge of sexuality and sexual health is based on where you are born. If you are born in a state that is abstinence-only until marriage, you'll likely get a textbook that is not only medically inaccurate, it may be accompanied by moral lessons that are sexist, homophobic, or just plain wrong. You might get a nurse who teaches that masturbation will lead to erectile dysfunction. In the past two decades, the federal government has spent two billion dollars teaching students information that is not evidence-based. As a teen in a conservative suburb in California, I was given a virginity flower demonstration during Sunday School which generally goes something like this:

A teacher shows the class a fresh flower bud. They pass it around, letting the class admire it. The flower, they tell you, is your virginity.

Then, they take the flower back. They crush and bruise and mangle it. "Who would want this flower now?" they ask.

Once you have sex outside of marriage, they tell you, you are ruined sexually. There's no going back. No one will want that flower.

Over the years, I've heard multiple versions of this demonstration. There's a piece of gum that the teacher chews up and then offers back to the students. There's a white sneaker version that becomes trampled and dirty. There's one where the students pass around a Snickers bar until it's melted and gross. They all end the same: No one will want you after you've had sex. More specifically, your future husband—because this is often directed to girls—will be disappointed.

I often think about how I felt limited growing up in a Catholic family, but at least I had access to San Francisco and its progressive resources. There are states, like California, that now have comprehensive sex ed including mandated LGBTQ education, but most do not. This is unacceptable. If I'd been born in Oklahoma or Texas or places without those educational opportunities, where would I be now?

There are so many passionate, talented people who believe in sexual wellness in the public sector, but our states limit what they can teach about sex, or even the questions they can answer. In many schools, a teacher who was approached by the young student I met in Kansas would not be able to discuss the topic with him at all, or could face sanctions if they did.

Even when they are able to talk, there aren't many resources. The Federal government has stripped funding for nearly all sex education besides abstinence-only. Their goal? To push sex out of the public sphere—out of schools, out of research universities, out of clinics, out of public health—to the private sphere, where it can be guided by conservative, religious institutions.

Sexual health advocates want better public sex ed, as do I. But as sextech entrepreneurs, we know that until the laws change, our best opportunity for educating the next generation may be in the private sector.

THE PATH TO ABSTINENCE

We currently have lower levels of sex education in America than we did twenty years ago.

Back in 1995, during the Clinton administration, 80 percent of students learned about birth control in schools.

But that changed in 2000, with the election of George W. Bush and the elevation of the evangelical movement. Today, fewer than half of the schools in the US teach any sex ed at all, and 75 percent of those that do, only teach abstinence-only programs.[22]

That means the generation of students entering college today haven't been taught about birth control or STIs, let alone more controversial issues like masturbation, sexuality, or consent.

In fact, nearly twenty states require that educators teach students that sex is acceptable only within the context of marriage.[23] Seven states prohibit teachers—under penalty of law—from even talking about gay people, unless it's to condemn them.[24]

The sad thing is, we know these programs don't work. The CDC and National Institutes of Health say that abstinence-only programs increase teen pregnancy and the rate of STIs, and yet—under both President Bush and President Trump, those are the only programs that get funded. (The current administration slashed more than

---

22   More on sex ed in America (or the lack of it) in my *New York Times* opinion piece, "How to Make Sex More Dangerous," March 2019.

23   State Laws and Policies, "Sex and HIV Education," Guttmacher Institute, August 1, 2019.

24   Research Brief, "Laws that Prohibit the "Promotion of Homosexuality": Impacts and Implications," GLSEN, Accessed online, August 2019.

$200 million from teen pregnancy prevention programs that had been funded during the Obama administration.)

Is it any wonder that we're at an all-time high rate of STIs?[25] At the federal, state, and local levels, there has been a sustained assault on sex education in schools, and yet the government still actively pushes abstinence and fights comprehensive sex ed.[26]

Within the government-backed school system, we have to fight for even the most basic information. No one talks about pleasure. No one talks about consent. No one talks about masturbation. How can we expect them to answer the types of questions I get from first-year college students?

For all the public conversation about consent and sexuality, we have generations of Americans completely unequipped to advocate for their bodily autonomy, that are extremely ashamed about any sexuality that they've experienced. We've failed those generations of women when we set them up to be hurt, and we failed those generations of men when we fed them toxic masculinity instead of teaching them about consent and pleasure for all bodies.

25 Ibid

26 Megan Donovan, "The Looming Threat to Sex Education: A Resurgence of Federal Funding for Abstinence-Only Programs?" Guttmacher Institute, March 2017.

Without an accurate education and healthy perception of sexuality, it's no wonder sexual wellness has so few entrepreneurs, fewer investors, and a market that doesn't quite know what to do with itself.

Imagine what our financial system would be like if we didn't teach math.

WHAT HAPPENED TO FREE LOVE?

When I talk with people from older generations about O.school, they're often confused. After all, they tell me—"we fought for this fifty years ago." They had consciousness-raising sessions and marches and "free love." Today, there's much freer discussion of sex on television, gay marriage is legal, and women are speaking out about #MeToo. How can it be that people coming of age now are still in the dark?

Most don't realize that as they've gotten older, they've lost touch with what students are being taught. They confuse conversation with education.

In many cases, they've also segmented themselves into progressive bubbles—like San Francisco, a bastion of "free love." It can be hard for them to understand that in other states, or rural areas, generations are coming of age, getting married, and having children with much less information about sex than they had.

Many may remember the freedom granted by the availability of birth control in the 1960s, which decoupled sex for pleasure and sex for reproduction, and allowed women to have sex without fear of pregnancy. They may remember the striking down of adultery laws, and the freedom to have sex outside of marriage.

These were, of course, huge accomplishments.

For heterosexual women born between 1933 and 1942, 93 percent reported having sex for the first time when they got married. About twenty years later, between 1963 and 1974, only 36 percent of heterosexual women reported waiting until marriage to have sex for the first time; the majority of women didn't wait.[27]

The Summer of Love birthed a generation empowered to take on its own sexuality, but we never educated them about it. They were left—like generations before them, to figure it out for themselves. The '60s and '70s birthed pioneers like Joani Blank, who started Good Vibrations; the Boston Women's Health Book Collective, which published Our Bodies Ourselves; Carol Queen, who started

---

27   Brian Alexander, "Free Love: Was There a Price to Pay?" NBCNews, MSNBC.com, June 2007.

the Center for Sex and Culture; and Susie Bright, better known as Susie Sexpert, who co-founded On Our Backs.

These pioneers—mostly women and queer people—helped change the conversation through grassroots educating. Many of us can thank them for our own personal revelations. (I know I can.)

But sex and sexual pleasure education have rarely come from institutions—it's too difficult to fund and too vulnerable to political change. So when I look to the future, I look to pioneers like them, and try to imagine what they would do if they were millennials. What could they do to understand the system and prove to the tech bros and banks and legislators that sexual wellness isn't just the right thing to do, but that it's a huge untapped market.

That's what sextech is.

THE LIMITS OF NONPROFIT

There are, of course, places other than schools where people learn about sex.

Planned Parenthood is the behemoth that gets the most attention, but it isn't alone. We're lucky to have organizations like Siecus, Advocates for Youth, the Healthy Teen

Network, and Power to Decide, as well as countless other smaller organizations doing community-based work.

But because so many depend on government funding, most focus on teen pregnancy prevention or STI reduction.

And with teen pregnancy prevention funds being cut, many nonprofits have had to refocus their efforts on advocacy rather than education. At the same time as funding has been cut, we've seen a sustained attack on reproductive rights across the board—many organizations are facing a war on two fronts.

Take Planned Parenthood. While it's a health provider at its core, it also has grant-funded sex ed programs—Planned Parenthood Online is one of the best providers of sex ed online today. But because of their government funding, and because they provide reproductive rights services, like birth control and abortion, they are vulnerable to the whims of legislators. Everything they do is under the microscope.

In 2014, anti-abortion activist Lila Rose went to a Denver Planned Parenthood Clinic posing as a fifteen-year-old asking about BDSM. The healthcare provider spoke to her about safe BDSM play and resources, unaware that she was secretly being filmed. When a selectively edited video was released to conservative media, it spurred the Colorado Attorney General to investigate the organization.

Even when private funders like Hewlett Packard and the Ford Foundation back organizations, their scope is limited—and further limited by the cautious nature of the large foundations that back them. Continued funding often depends on a measurable bottom line—stats like teen pregnancy and STI reduction—rather than broader education about consent and wellness. As much as educators might like to talk about consent and pleasure, their mission statements don't allow it in any measurable way.

Nonprofit and social impact organizations who want to focus on pleasure as part of comprehensive sex education or part of an activist cause often must rely on crowdfunding and community support.

LESS ROOM FOR RISK MEANS IT'S ALWAYS A SIDE HUSTLE

But it's not just conservative funders or limited missions—in many of these organizations there are massive disincentives for risk-taking.

In the startup world, you're supposed to take risks. A venture fund wants to make a profit, and as such encourages you to find the best people and understands that often, you must compensate them competitively. It's a risk, sure, but it's one investors are willing to take for such a large payoff. Nonprofits, which depend on incremental success for continued funding, are berated for any increases in "overhead"—how much you pay your staff. We expect the people who are trying to make the world a better place to do so for almost no money.

That's not to say these people don't want to take risks. I once met with a large organization in Los Angeles that provided healthcare, housing, counseling, and a host of other services for the LGBTQ community. We sat for hours talking about the innovative and potentially program-altering changes that O.school could bring. The conversation was electric—there were so many possibilities that the staff raised.

But at the end of the conversation, they conceded that there wasn't really a way forward. Their funders depended on very specific metrics, and they were already so stretched for services that they couldn't afford to take risks when that money could be used to keep a trans teen off the streets, or provide HIV meds for an elderly gay man on disability.

I understood completely—but it made me that much more dedicated to making a space where risks could be funded. Where there is room to experiment with new ideas.

When you're a private company funded by venture capital, there are stages. The more money you raise, the more that you have to prove. In the pre-seed stage, it's understood that you've only got an idea and you're allowed to not have things figured out. By the time you get to the next stage, Series A funding, you'll need to be able to really show measurable traction. There's still room to experiment and change, of course, but investors start wanting to see results.

There's very little risk capital available in the nonprofit space. For a nonprofit to get the amount of grant money a startup might get from a pre-seed round, it would have to provide Series A level metrics. That means most new nonprofit organizations merely follow the path of existing successful programs, rather than striking out on their own.

Nonprofits have to prove so much more before they can raise a dime.

Nonprofits have access to capital but aren't allowed to take risks. On the other hand, sexual wellness entrepreneurs can take risks, but have very little capital. We can't get nonprofit grants, and many big investment funds are wary of us. (I'll talk more about this in the next chapter.)

There have been success stories, for sure—but for the

most part, sexual wellness and sextech entrepreneurs have been individualists that use crowdfunding or personal equity to start businesses that they hope will make it to sustainability, if not runaway success. They are content creators, online educators, sex therapists, or sex toy designers. Their ideas may be brilliant, but they have to fight for every bit of growth—like an ambitious plant on the side of the rock face of a cliff.

The lack of money diverted to us often means no one— even those with tremendous ideas—can get to scale. As a result, most founders and creators, no matter how ambitious and innovative, stick with a nonprofit mindset, always dying for money, doing the work for nearly free, because they believe what they're doing is important and vital—what other option do they have? Many devote their lives to it, but usually as a second or third job. Talk to most sexual wellness entrepreneurs and you'll hear the same thing. It's like art, music, activism. You can't not do it—it means too much to you.

As long as so little funding goes to these sectors, this is where we'll stagnate, with everyone scrambling for scraps and hoping it's enough to keep their ideas alive.

But what would it mean to think about sexual wellness on a huge scale? What would a legitimized sexual wellness space look like?

THE PRIVATE SECTOR IS ESSENTIAL

The market for sexual wellness is like women's sexual pleasure in general—the world knows it exists, but doesn't like talking about it, doesn't really know how to achieve it, and is uncomfortable investing in it.

Our schools are failing to teach even basic anatomy and reproduction, let alone pleasure. Our public health organizations are stretched for even basic funding, and our corporate nonprofits are so focused on narrow goals that even innovation risks loss of funding. And with conservative legislators ascending, it's only getting worse.

And yet, there are eight billion people who are radically underserved in terms of sexual wellness. The toll, physically, emotionally, and socially is immeasurable.

We are slowly growing more and more aware of the costs, individually and collectively, but the shift we want won't happen without the private sector.

The activists, thinkers, and movements throughout have made a huge impact—but until we connect them with private sector investment, until we find a way to reward and finance them, until we connect what they do to a market, their impact will remain stunted. As an entrepreneur, I believe that the risk-tolerant culture of tech has the ability to effect widespread massive cultural change in one generation.

I don't want to wait until the government permits us to learn about consent and pleasure—if we're not all dead by then, the sexual wellness potential of countless lives will have been squandered. There's no appetite for a government-funded program to address sexual wellness, and structurally it's near impossible to execute. Imagine if we could do with sexual wellness what Twitter did for political activism.

To move quickly, we need capital. To get capital, we need support for private sector innovators and to prove the viability of the sexual wellness market. And we owe it to the next generation to move quickly.

CHAPTER 4

WHO BUILT THE INTERNET?

I'll never forget the moment I realized the true scope of this problem.

I'm at one of my first investor meetings with one of the funders of my previous company. He knows my track record. He knows I'm serious. Over sparkling water at the St. Regis, I make my pitch for O.school—I explain the near total void of sex and pleasure education, especially for women. A market without a dominant voice. A $27B retail segment that major tech companies like Facebook and Amazon probably won't—or can't—touch.

As I go through the deck, he blushes at first, then looks confused, and then...pity. He sort of half-smiles, patronizingly, prepared to drop a major truth bomb on me.

"Andrea," he says, "I'm confused why you plan on focusing on educating women? If you want women to have more pleasure, then you should educate men. After all, men are the ones who give women pleasure."

There are biological realities of the world, he continues to explain. Women aren't as biologically interested in sex as men. Any internet space like O.school would just degrade to porn.

I stay calm. He does lay out some core truths, but they're not what he thinks. Rather than dissuade me, in my head he's proving my point. People like him are the very reason I needed to build O.school. I end the meeting as quickly as I can.

I laugh about it now—the fact that one of the men who helped fund major internet companies could have such outdated views of sexuality. But it's no laughing matter. Men and like-minded organizations continue to be the gatekeepers of funding for sextech, without understanding the basics of sex.

Sextech founders balance all of the immense struggles all founders face, but with the added burden of stigma. It's one of the last frontier markets, and few people outside our community acknowledge that. Many of us feel unhinged from the dissonance of experiencing the demand for our products every day with everyone we meet—from strangers overhearing conversations at coffee shops, Lyft drivers, and even the investors themselves—only to be passed over again and again by investors.

I recall being held up for fifteen minutes in a women's bathroom by a partner of a well-known investment fund, who had just had a baby and wanted to talk about her current difficulties having sex with her husband. That fund also passed based on the "lack of opportunity" they saw in the blue ocean $122B projected market opportunity we educated them about.

If we can't trust investors to be greedy, what can we trust?

We don't have the old boys' club to pass down knowledge or connections. We don't have an established market or a course at Stanford. We're a massive multibillion-dollar market, and we barely have a Wikipedia page.

We need to fix that.

In this chapter, I'm going to lay out the structural problems we face in funding, as well as the double-edged sword of stigma and sexuality in today's market. Much later in the book, I've included a quick how-to guide for anyone willing to join us in tackling these massive, man-made problems.

WHAT GIVES, PRIVATE SECTOR?

Most founders have the same tools available to them. The same payment processors. The same website builders. The same tech conferences and advisory boards and venture funds. Silicon Valley is filled with startups, all driving down the same stretch of 280, often looking to be bought by the same publicly traded companies. It's the mainstream ecosystem.

But try to do something in sextech—or even the most anodyne corners of sexual wellness—and you'll suddenly see barriers that you never knew existed. Moral clauses buried in Limited Partner agreements. Banks who will shut you down and tell you to take your money elsewhere. Payment processors that charge exorbitant rates. Social media platforms that won't let you advertise. And all the while, you'll likely find well-meaning mentors and investors who will give you advice for hours and gladly go to endless coffees and dinners (especially to address their own personal problems)—but won't write checks.

Sextech isn't the only market that's denied access to this ecosystem. Startups that work in cannabis and cryptocurrency are similarly locked out of traditional banking and advertising. I don't have anything against those categories, but in some ways you can see the rationale—under Federal law some of what they're doing is still illegal.

Sextech and sexual wellness are not illegal. And yet, the restrictions can be even more aggressive. People love to talk about blockchain and cannabis. Despite the dangers, investors can at least understand these markets.

So why hasn't the sextech revolution happened yet? I can tell you the answer:

Money.

Despite the size of the total market, there's been no unicorn for sextech. There's been no "a-ha" moment for investors. Until this happens, there's no incentive for these gatekeeper companies—or investors—to break down the doors.

We're going to do it. But in order to do that, we need to unpack the reasons those doors are closed. I need to take you several steps back in the process, and show you where the real money is, and how the decisions are made.

Because if the person saying no is a well-meaning investor or the founder of one of those NASDAQ elephants, the blockage is more structural than personal.

WHAT IS VENTURE CAPITAL?

Esteemed professor, startup expert, serial founder, and

author of *The Four Steps to the Epiphany*, Steve Blank, defines a startup as a "temporary organization in search of a business model."[28]

Venture capital is the way that tech startups fund this search. Entrepreneurs engage in rapid experimentation until they find out how to make money, essentially buying themselves time until they unlock and create a business model that didn't exist before. When a startup finds its business model and establishes monetization, it becomes a company. Until that time, they need venture capital to survive throughout their idea (or "seed") stage—and then more capital to grow and capture market share.

Venture capital is a bizarre concept if you aren't familiar with it. As a founder, you're given a pile of money to experiment and to try to build a dominant business model. It's not a bank loan that you have to pay back. Venture capital is, and should be, risk-friendly capital—the same capital that nonprofit organizations do not have access to, that we covered in the last chapter. About 75 percent of all venture-backed startups fail, and early stage investors expect to lose their money most of the time.[29] They understand that in order to maximize

---

28 Steve Blank, "What's A Startup? First Principles," Steveblank.com, January 2010. Referenced by JG Vargas-Hernandez, "Development of Entrepreneurship in ZMG: Growth Startups and Precursor of Innovation and Economic Development," *Business and Economics Journal*, 8: 277. doi: 10.4172/2151-6219.1000277.

29 Faisal Hoque, "Why Most Venture-Backed Companies Fail," *Fast Company*, December 2012.

success—and return on their investment—we need an ecosystem in which the vast majority of ventures fail.

Rather than demanding small returns on each company, they are betting that one of their investments will be a billion-dollar or more win, often called "unicorns." These unicorns form the lifeblood of the venture capital industry. One "Uber" or "Facebook" is all they are looking for. Venture capital firms may have different value systems, also referred to as "theses"—but this big payoff strategy is the mindset of venture capital.

There are only four outcomes to every startup venture: death, IPO (the "initial public offering"—which means raising funds in a public market like a stock exchange), acquisition by another company (say, Facebook or Google), or operating as a self-sustaining business. The last one is not particularly favorable to investors, since it scales slowly, and the capital they invested is stuck in the business.

So the key to understanding investors? Maximize greed and reduce risk. However, as massive as the sextech opportunity is, greed has not motivated the best venture capital funds to jump on it. Rather, it has remained underfunded until even 2019. Let's dive into the decision-makers to understand how this could possibly be.

LIMITED PARTNERS: THE FUNDERS AT THE TOP

Limited partners have a major impact behind what gets funding, what doesn't, and the internet ecosystem we have today.

When a startup founder fundraises, they raise money from investors, like angel investors, who invest their own money, and venture capitalists—also known as VCs—who invest other people's money. VCs make money themselves when they make other people money. Those other people—the ones with the money to invest—are called Limited Partners, or LPs. LPs can be wealthy individuals, other venture funds, a family foundation, or they can be large, institutional funds. Typically, the larger the fund, the bigger the check. Eligible institutions can be huge, from the Vatican to venture arms of big corporations to even countries. LPs set the ground rules.

Occasionally, you'll meet with a VC who is also an LP—perhaps it's a small fund, or they take a more hands-on role. But mostly, LPs are behind the scenes.

What LPs believe—financially, strategically, politically, morally—matters for the rest of us. They can influence what a VC can fund, from the sector of the market, to the risk profile of the startup.

While LPs come in all shapes and sizes, for the most

part, they're looking for a return on investment. They may have broader ideas about how they want the world to look, or what types of startups they want to fund. There are LPs and smaller funds, for instance, that focus on gender diversity. But even those with a social message are not to be confused with charities.

The DNA of LPs gets passed down to VCs, and ultimately down to the founders of the startups they invest in. So while much attention has been paid to diversity in the workforce, until it's a priority of the overwhelmingly white, overwhelming male LPs, founders won't feel much pressure to diversify their own companies.

Most LPs want to fund companies with the potential to become the new Facebook, Google, or Uber. They want companies with the potential to define a category.

More important than the LPs themselves are the LPAs— "the Limited Partner Agreements"—the agreements LPs sign with the funds into which they plan to invest. Many of these agreements can have clauses that specify investment sectors that are off limits—most commonly guns, drugs, and, unsurprisingly, "adult" businesses. This was originally meant to keep funds from investing in porn. But as I learned early, LPAs regard "adult" as a slippery slope—"adult" has come to mean anything that involves sex.

Given the vagueness of "adult," these restrictions can be enforced inconsistently. While investors in tech are not generally risk averse—most plan to lose on most investments, but make it up with a few outsized returns—it's simply not worth it to violate the LPA. But even without a legal LPA in place, or one that prohibits "adult" investments, all it takes is a concerned phone call from another partner, or a negative reaction in an LP meeting, and the deal is dead.

VCS AND INCENTIVES: WHO WATCHES THE WATCHMEN?

Venture capitalists are the next rung down from LPs, and are the ones who most often sign off on the investments. They have to answer to the LPs, ultimately, but are largely empowered to go out and source potential investments and make the deals.

Like LPs, VCs are overwhelmingly white and male. Want proof? Look at the team page of any VC firm's website. Patagonia vests galore.

Fewer than 10 percent of all decision-makers in venture capital are women. And nearly three out of four VC firms in the United States have no women investors at all.[30] Zero.

---

30 Kate Clark, "Female founders have brought in just 2.2% of US VC this year (yes, again)," TechCrunch, November 2018.

VCs focus on the probability that the business will generate an outsized return to the fund, based on pattern matching of founders and rate of growth. Investors especially love the "stickiness" of a product—how 'addicted' someone can get to a product. Think of how many times you scroll Instagram, or how much time you spend on Netflix.

VCs back founders who "match a pattern" that might indicate they'll be the next Mark Zuckerberg. For example, that pattern might be a Stanford pedigree, or a history of smaller, successful companies...or someone who "seems" like a founder. Think of all the CEOs in hoodies or black turtlenecks, and you'll get a sense of what signals they send to investors.

Of course, this creates a huge bias against founders and businesses who don't fit this pattern. Women. People of color. Older people. Queer people. Parents.

Numerous studies have found that teams with at least one female founder perform better than teams with an all-male makeup, but little has changed. In 2018, female founders brought in only 2.2 percent of all venture capital that year. Women of color got an even tinier fraction.[31]

---

31   Nina Zipkin, "Out of $85 Billion in VC Funding Last Year, Only 2.2 Percent Went to Female Founders. And Every Year, Women of Color Get Less Than 1 Percent of Total Funding," *Entrepreneur*, December 2018.

Not only is this bad for those of us who don't match the profile of what a founder "looks" like, it's bad for investment. VCs talk a big game about finding untapped markets, discovering explosive financial potential, but when it comes to their pattern-matching predilection, it can produce cookie-cutter founders.

Confirmation bias in their pattern matching and the limited lived experiences of a VC can block innovation in entire sectors.

It also means they miss major investment opportunities, like the breast pump. The breast pump, which extracts breast milk for later feeding, sucked for generations. While the technology was invented in 1854, it wasn't until women began entering the workforce in earnest in the 1960s that it became an essential component of motherhood. But comfort, convenience, speed, and dignity weren't really concerns. Neither was the tech.

Major investors weren't interested in the market. Maybe it was because we have a fantasy of what constitutes traditional motherhood. Maybe it's because investors didn't view it as a real problem—at least not one they've experienced. Markets ignored it, because women were ignored.

The situation was so bleak that in 2014, women even

came together for a "Breast Pump Hackathon"—desperate to find some tools that would give them freedom.

It took a woman-led startup, Elvie, to take the matter seriously. Today, with Elvie's wearable breast pump, the women's health company announced it had raised $42 million dollars. CNET said the pump "feels like breaking out of jail."

Babies weren't invented in 2019, of course, or even 2014. Billions of people have struggled with breastfeeding—it just took investors and innovators until now to start taking the problem seriously. If more people who lived through breastfeeding with a crappy pump sat at VC tables, this would have been solved long ago.

For a VC to invest in a company, they need to comprehend the problem and the market—not just as numbers on a spreadsheet, but like a lightbulb going off. They love nothing more than to have a founder pitch them with an analogy to an existing "unicorn" or billion-dollar company—an Uber for cleaning! The Airbnb for camping. Tinder for mating cattle. (All of these exist now.)

When it comes to modern sextech, sexual wellness, and femtech, much of the market is led by women. Most of the founders are women. But most of the VCs are men. It's hard for VCs to take problems seriously when they've never experienced them. "How hard can it be to pump breast milk?" they wonder. They look on Amazon and see a market with plenty of products. In many cases, they don't see the potential, because it doesn't affect them.

Meanwhile, during the early 2010s, there were a plethora of laundry startups. Why? The biggest problem in a young Stanford graduate's life is hating laundry. Regardless of inanity, it was something that founders and VCs could understand.

This doesn't mean we shouldn't apply scrutiny to the founders of companies—but who and what gets funded starts with the VCs and LPs.

A more diverse venture capital ecosystem, with a broader range of investors and lived experience, could be the solution to reaching these other markets. But that takes time.

Arlan Hamilton, the founder and visionary behind Backstage Capital, has dedicated her career to giving access to capital to underrepresented, or who she calls, underestimated founders. There are more diverse funds, but the issue is that most of these funds are relatively young, and still struggle to raise money from LPs. Just as interest compounds on capital, the funds that were established early in the internet have compounded their reach and experience. For some, there hasn't been a sufficient amount of time for their investments to prove themselves, and they struggle to raise from LPs without the success that an older, more established fund might have.

In other cases, they have different values beyond maximizing returns. This is great for society, but can cause large, established LPs to look elsewhere. LPs more often go for the VCs who can maximize their returns—and thus look for those with proven track records.

Maximizing returns is an understandable goal, and it's driven much of the internet so far. There are, however, other metrics that we should think about applying. For example, what would tech look like if VCs valued the effect a company might have on society? It's a longer-term view, but one that arguably has a higher payoff.

THE DIVERSITY DILEMMA MIRRORED ONLINE

Could Facebook have avoided the democracy damaging flood of fake news, or the siloing of communities? Could YouTube have started addressing abusive comments, or stopped spurring the explosive growth of conspiracy theories? When your goal is keeping someone engaged as long as possible, you tend to feed them what they want to have, what makes them feel good, what bolsters their own opinions—rather than what is actually valuable.

Twitter is one of the best examples of how lack of diversity plays out online. For many of the cis white men who founded, invested in, and ran Twitter, it is a bastion of untrammeled free speech. A marketplace of competing ideas. There's a strong theoretical argument for platforms to be free spaces, blind to race or gender or sexuality or ideology, but in reality it doesn't always mean equal access.

As a queer woman of color, I knew how uncomfortable online spaces could be. One of the reasons I started O.school was because like so many women and gender non-conforming people, I had been harassed online. At times, going online wasn't so much a free exchange of ideas, as much as target practice for racists, misogynists, and homophobes. For many people, merely offering up an opinion results in comments about their looks, their weight, their intelligence, their reproductive capacity, their race. It means being bombarded with rape threats, death threats, and other harassment.

I'm not saying that white cis men don't get attacked, but it's rarely the swarm it is for women and minorities. A staggering 81 percent of women experience sexual harassment in their lifetime, and the internet isn't any safer than the offline world.[32] Of the 40 percent of Amer-

---

32 Rhitu Chatterjee, "A New Survey Finds 81 Percent Of Women Have Experienced Sexual Harassment," NPR, February 2018.

icans who have experienced public bullying, trolling, and harassment online, women are twice as likely to experience gender-based harassment than men.[33] Without being in the decision-making spaces, we can't contribute to making those platforms safer and more inclusive.

What if, instead of maximizing "stickiness," Twitter had engaged with social scientists? What if their executive staff were more diverse? Could they have seen this coming? Could they have built in better moderation and allowed less toxic behavior? Jack Dorsey recently talked about how beneficial for the platform toxic behavior is—and is now thinking about how to change Twitter's incentives.

VCs love addictive systems. Hooked by Nir Eyal, one of the most influential books in tech, explains how apps on our phones keep us engaged—and the effect it has on users. More importantly, it details how to accomplish this. VCs use the potential 'addictiveness' to evaluate startups, without much concern as to whether those effects are positive or negative for the end user, let alone society. No matter what aspect of tech, most potential investors will ask you for these metrics—stats like daily active users, time spent in the app, retention rates after three, seven, or thirty days. If you can keep someone hooked, you can

---

33   Maeve Duggan, "Online Harassment 2017," PEW Research, July 2017.

likely get funded. In Silicon Valley, addictiveness is a virtue.

This drive for stickiness imbues the whole ecosystem. Even if a founder felt concerned about the effects of their product, they would have to keep in mind the opinions of their investors, lest an investor stop funding them.

For new companies, it can often feel like a deal with the devil. VCs are relentless about growth. If you're not meeting those metrics, you risk losing your funding, and your company dies. It's not enough to grow. VCs push startups to grow extremely fast. Twitter was a classic example of this. The team thought they were building a bottle rocket—then watched that rocket to go to the moon as popularity surged, and investors pushed for more. The social media directive has been growth over everything; pausing to understand how that growth might affect users, or whether safeguards might be built in, aren't even secondary concerns. They are barely concerns at all.

Most of the social media platforms we use today grew too quickly, with too few experts or social scientists. Founders have an idea—an experiment—but by the end of the process they're more like mad scientists. It's no longer about finding out what the experiment will produce, as much as it is constantly manipulating it to produce the results the VCs want.

This is a big failure of modern capitalism in Silicon Valley. On one hand, the process lets people take risks, allows entrepreneurs to innovate at a massive rate, and empowers visionaries to build at an unforeseen scale. On the other hand, it leaves havoc in its wake. We measure the edifice by the height of the spire, not by the cost of the build. We only look at the profit, at the capitalization, at the growth, without looking at the costs to society or the people using it.

I don't believe most tech founders set out to hobble democracy or foster harassment or any of the other social negatives that have come in their wake. I believe that the tech ecosystem and incentives pushed them at breakneck speed, and created the internet that we have now.

Much has been written about the need for more diversity and inclusion in the tech world. Emily Chang's *Brotopia* is a searing review of Silicon Valley today and talks about the issues that arise when we don't include stakeholders.

Many organizations, such as Project Include, Change Catalyst, Tech Inclusion, and others—have invested billions to diversify tech, but the reality is that, online as it is offline, the structural bias of tech is massive and was formed long ago.

THE DNA OF THE INTERNET

There's a dictum in Silicon Valley that founders should identify a problem in their own lives, and design the solution. Silicon Valley has largely been the domain of straight, white men—it's not surprising that what they built solved the problems effectively for them.

Take Facebook. Mark Zuckerberg first designed the platform as a way to more effectively rate the attractiveness of women at Harvard. As it grew, it became a way to find out more information about these women and his other classmates. While its scope has grown, many of us still use it that same way—to snoop on ex-lovers, family members, potential dates, employees. Companies—from Apple to Cambridge Analytica—use it the same way. That feature is built into its DNA.

It works terrifically well for straight white men like Zuckerberg. Its addictive nature compels people to post highly personal information. But straight white men are less likely to face harassment online. They're less likely to have to deal with stalkers. They don't need to worry about being "outed" as gay. They are less likely to be surveilled by law enforcement.

Founders shape the DNA of their entire organization. When you look at the DNA of some of the most toxic, prominent spaces on the internet—Twitch, Reddit, Twitter, YouTube, Facebook—few were designed with women, people of color, queer people, or people with different lived experiences.

At first glance, practices and policies that these massive platforms have created sound agreeable. But without diverse lived experience, the picture changes. Think about Facebook's real name policy, which originally mandated that one's profile name match your government issued ID.

A straight white cis male would have little reason to think "real names" might be an issue—in fact, he'd be likely to feel it is a positive feature. The value of information and verification trumps privacy. But for a trans person, that's not necessarily true—a government ID isn't necessarily reflective of their lived identity. For domestic abuse or stalking victims, the insistence on real names leaves them vulnerable to further harassment and attack.

The enforcement of the policy in 2015 led to widespread protests, and Facebook eventually initiated a compromise to protect those in special situations. But that same DNA still informs countless other aspects of the company.

When I designed O.school, it was important to me to build a platform that would be responsive to my own DNA, my own lived experiences. We consciously did not require real names, or any names, to participate in conversations. Every livestream class we did came with a live moderator to prevent harassment and abuse, and we worked to find tech that would supplement this in other areas.

As we've discussed, sex education is abysmal in many places, which has sent many of us online to seek out information. In many ways, the internet has been incredible for those of us who are non-normative, sexually or in gender presentation. It's allowed us to connect and build communities and find resources.

Unfortunately, it's also a minefield. On most social media platforms, women and gender non-conforming folks face routine harassment, especially when looking for information on sex. And unlike information on, say, woodworking, the information on sex and sexuality is incredibly unreliable. Mixed in with the good stuff is schoolyard misinformation, conservative dogma, and sexually explicit content.

The internet has evolved to solve so many problems... except this one.

THE (NOT SO) SECRET DRIVER OF TECH INNOVATION

The first generation of people on the internet had a vision

of connecting people together to create communities. The pioneers and early adopters of the internet tended to be men, often solitary ones who didn't always feel connected to a larger society. Maybe you were the only kid who liked Dungeons & Dragons at your school. With the internet, you could quickly find other gamers like yourself online.

They also tended to be younger men, kids in high school and college with access to a personal computer and a modem connection. Gaming and social connections may have been the initial driver of internet adoption, but sex— or more specifically, porn—quickly replaced it.

Porn has long been one of the chief drivers of technological innovation. New technology begins as the domain of experts and professionals—but the ability to access sexually explicit material is what brings in the masses.

The production and consumption of sexually explicit material drove the adoption of Super 8mm film in the 1950s and '60s, Polaroid cameras in the '70s, and the VCR in the '80s. In the '90s, it was a Dutch porn company, Red Light District, that pioneered the first workable video streaming system. Porn raised demand for internet speeds and propelled peer-to-peer sharing.

Even the boring tech-affiliate payments that most companies use or payment processing systems that allow you to pay online with a credit card were either birthed by or popularized by those who created porn, or those who wanted to access it. Sex is what gives tech its "stickiness."

It's funny—and infuriating—that adult topics are barred from so many online spaces without any respect for these topics having led the innovation in the first place.

Like Facebook, Twitter, and much of the rest of the internet, online porn was designed with men in mind. A founder of a sex toy company that's been around since the '70s told me that, historically, only 2 percent of all porn purchased was purchased by women. While those numbers may be changing, companies build for their paying audience. Porn, which built the internet and drives development, is built by men and for men.

While LPs and VCs refused to touch sexual wellness or sextech markets, a second ecosystem flourished in the adult world funded by private investors, led by one company that very few people have heard about.

THE BIGGEST MONOPOLY YOU'VE NEVER HEARD OF

When people talk about dominant tech behemoths, we all hear about Facebook, Google, Netflix, and Amazon. But have you heard about MindGeek? Most people have not. Yet, they compete at the same level of influence online, with only a fraction of the public attention.

Amazon has been, and continues to be, the killer of smaller e-commerce platforms. The same thing happened with the emergence of MindGeek, which consolidated the market and made starting an independent premium porn company much less viable.

Porn used to be a massive, lucrative business. It's still lucrative, but only for very few companies, with this one emerging as the most dominant.

Most people can recognize Pornhub, the number one free tube site for pornography, which is part of a near monopoly on porn. It and eight of the top ten porn sites are currently owned by one company—MindGeek.

The stats on this monopoly are incredible. One out of every six people in the world hits a MindGeek site every day. There are claims of over 100 million daily visitors. Some report that they consume the third largest amount of bandwidth, with only Google and Netflix ahead of them. [34] [35]

---

34   MindGeek, Accessed June 2019. https://www.mindgeek.com.

35   David Auerbach, "Vampire Porn," *Slate*, October 2014.

Originally called Manwin, MindGeek was built by a guy named Fabian Thylmann, the father of modern porn. It was launched just a year and a half after YouTube, and a month prior to the arrival of the iPhone. Its business model was influenced by the former, but its massive success hinged on the latter—profiting from millions of user-uploaded, often pirated videos, available in the palm of your hand.

The effect on the adult industry was immediate. Companies, already struggling with piracy via torrent sites, would see their newly released scenes online often within days. More importantly, consumers could access them without venturing onto torrent sites, or worrying about keeping downloads of their files. Financially, the bottom fell out of production, studios collapsed, and wages for performers fell. What had been a fairly diverse ecosystem of independent producers began to struggle for oxygen.

Flush with cash from his venture—and $362 million of debt equity investment in the form of a loan by private equity firm Colbeck Capital—Thylmann started buying up smaller tube sites and distressed production companies. He started with Mansef and Interhub, then the original owners of Pornhub and Brazzers, followed by nearly every porn company you've ever—or never—heard of: Reality Kings, RedTube, Videobash.com, EuroRevenue, Celebs.com, Twistys, GayTube, SexTube, Digital

Playground, xTube, and more. In over five years, they were able to secure over one hundred websites.

If it weren't porn, it might have been an antitrust issue.

The reach is incredible—and it's completely without representation. One company owns both production and distribution. In this scenario, consumers lose. The market isn't optimized for creating the best product possible, but rather the one that appeals to the broadest audience.

However, one effect was a new wave of platforms that would grow on the backs of MindGeek's consolidation—such as clip sites like Clips4Sale, fan subscription sites like OnlyFans, and live cam platforms like LiveJasmin and Chaturbate—which enable individual content creators to create income streams and reach customers directly, similar to Uber, Etsy, and the age of the gig economy.

It's possible that had MindGeek not come around, the adult industry could have been a fertile testing ground for sextech—a secure, sexually aware, pleasure-focused audience. But the existence of this one massive-traffic behemoth means that nearly everything in the adult industry is now geared toward the lowest common denominator—and mainly targeted at men. Ads for low-cost penis pills, sex video games, and dubious dating apps are much more profitable for MindGeek than ads for cutting-edge vibrators.

THE WORLD'S DE FACTO SEX EDUCATOR

The top ten porn sites receive upwards of 350 million visits a day, and three of the top ten sites in the US are porn.[36] The top US porn site—Pornhub—has more visitors in the US each day than Reddit, eBay, or Wikipedia, and it's not alone. It's the sixth most visited site in the country, and the eighth most visited in the world.[37]

PornHub and companies like it, known as 'tube' sites, took adult content from behind a paywall, and made it accessible to anyone with a smartphone. In less than ten years, they became the de facto source of sex education for a generation.

Now, I'm not against porn. Far from it. I think that, at its best, porn can be tremendously informative and educational, especially for those with non-normative sexualities or desires. It has the ability to help us understand our desire and our sexuality, to let us feel less alone. It can also be fun, arousing, and enhance our sex lives.

---

36   SimilarWeb, "Top Websites Ranking," Accessed online, August 2019. https://www.similarweb.com/top-websites/united-states.

37   MindGeek, "MindGeek by the Numbers," Accessed online, August 2019. https://www.mindgeek.com.

Too often, however, porn exists in a contextless void, in a world where sex education has been decimated; where abstinence-only sex talks from parents never touch on desire, consent, or pleasure; where misinformation online can bolster myths reflected in millions of online porn videos; and where only select tastes and preferences are centered.

We don't know the long-term or short-term effects on porn, because of the relative difficulty in studying it. Porn, like sex, is highly politicized.

Funding for the study of porn is extremely limited, even at the most basic research level. The federal government all but refuses to provide grants to study it—imagine the bellowed outrage in Congress—and private foundations have largely sidestepped the subject as well.

Thus, most of the studies that have been done have been done by those who believe porn is a societal negative. Even for the researcher who wants to remain neutral, it's hard to find a control group. Canadian researcher Simon Lajeunesse, a researcher who found that the average age young boys first search for porn is age ten—perhaps not coincidentally, the average age children get their first cell phone—had a harder time determining its effects.[38]

---

38   University of Montreal, 'Are the effects of pornography negligible?" *ScienceDaily*. l. (2009, December 1). Retrieved August 6, 2019.

"We started our research seeking men in their twenties who had never consumed pornography. We couldn't find any," he said.

Our country's educational, governmental, and often parental, approach to porn is similar to our approach to sex ed—abstinence-only. We cannot talk about it in schools, we cannot provide grants to study it, and parents barely have the tools to address sex and dating, let alone porn.

No one is talking to their kids about responsible consumption of porn. It's all done in the dark, unguided, without teaching them that porn is a fiction. The porn industry is not built for sex ed. Porn rarely displays consent negotiations or condoms. Most porn tends to focus on a specific type of body, while race is shockingly reductive, and women shockingly demure. There are unrealistic expectations about performance and genital size. Kids don't know how to navigate that. It's like letting them learn how to drive by watching *The Fast and the Furious* movies.

While projects like Boston's Start Strong have begun the discussion of teaching kids "Porn Literacy," it's not just kids who need it. We all do. It's one of the reasons I started O.school.

But if porn has been affecting our sex lives, and our society, it's done an even greater number on sextech and sexual wellness. Its prevalence has made it harder for companies to get funded. Many investors incorrectly assume that porn is profitable and therefore suspect sextech companies will be tempted to build sexually explicit content.

The prevalence of porn has meant that everyone, from Apple to Starbucks to the UK government, is creating filters to block it. Unfortunately, as I learned early on in O.school's development, those filters also block things like...O.school.

So when that investor told me that O.school would inevitably "degrade to porn," he wasn't just worried about explicit material, he was likely worried that we would be ghettoized. Lumped in with the perverts and the banned sites and the dildos hidden behind the counter.

It's a real fear. When Oregon-based tech pioneer Lora DiCarlo won an innovation award from CES—the high-profile Consumer Electronics Show that takes place in Las Vegas every year, and is covered by hundreds of media outlets—they were elated. Their body-responsive vibrator was poised to get global attention.

Until the board at CES realized what it was, and took the award back. As the prominence of porn companies like

MindGeek rose, CES pretty much banned adult content from its show floor. And with it, sextech companies like Lora DiCarlo.

The fear that discussions of sex are the same as porn—and thus, will lead to depravity—are deeply ingrained in us. Conservatives from Alabama to Moscow fight to stop people from discussing homosexuality because they fear that it will turn people gay. Others fear that talking about gender issues with children will make them trans. That talking about BDSM will unleash an epidemic of bondage.

And in some ways, they're right—each discussion of these topics does encourage people to be less closeted, and to be more open about the desires they already feel. It's not causing them to be gay, or to try bondage, but it does allow them to do it with less shame and more safety.

That openness is what drives sextech. And that fear is what has kept it from growing. The fear isn't just that a company like O.school will become porn, it's that by opening the door to sexual wellness, we're opening the door to porn. After all, in a society where we confuse cause and effect, where we worry that conversations about sex are the same as sex talk, and where these conversations cause instantaneous transformation, how are they supposed to tell the difference? Investors, corporations, and social media platforms are just externalizing our society's deep-seated fears.

Even crowdfunding sites like Kickstarter, Indiegogo, and Patreon—the platforms that made early sextech companies like Crave, Dame, and MysteryVibe viable well before most VCs would acknowledge the market—have struggled to balance approving sextech while keeping out pornography.

I'm glad there are those who are innovating and creating diverse expressions of sexuality in the porn and erotica world. There are amazing people, such as Erika Lust with XConfessions, Cindy Gallop with MakeLoveNot-Porn, Michelle Schaidman of Bellesa, and Jiz Lee and Shine Louise Houston of Pink and White Productions, creating high quality, diverse, ethical porn and erotica. In the audio space, Gina Gutierrez and Faye Keegan of Dipsea are innovating sexy audio stories made especially for women. We need to find a way to help independent voices in porn compete as well. It's part of the ecosystem of sexual wellness.

Sextech can bring the revolution that contextualizes porn, can help us have the important conversations around sex, and can help corporations—from Apple to Facebook to Patreon—understand the difference between sexually explicit content and age-appropriate educational material.

For too long, sextech has been treated as a slippery slope toward porn, by investors, platforms, filters, financial institutions, conferences, and legislators. As such, they use outdated, inaccurate, and inefficient ways to segregate sexual wellness, as a way of trying to keep porn out.

But here's the secret: the market is too big, and the world is changing too quickly. The internet built by white men isn't prepared for the revolution that's coming, and they're going to miss out on one of the last, great retail booms of the internet.

I don't resent the men who don't understand the necessity of sexual wellness, nor the ones who construct the platforms that exclude it. For me, it's another opportunity. When that investor told me that if I was going to improve women's sexual pleasure I was going to have to teach men, he wasn't exactly wrong—he just didn't know what I needed to teach them.

THE INTERNET NEEDS SEXUAL WELLNESS

One of my early investors, Jake Gibson, was also one of the co-founders of NerdWallet. When I asked him why he was investing in O.school, his answer was simple:

"When the internet started, it was a place for all freaks and geeks," he said. "It was a safe space for them in a way that other places in society weren't as safe. As the internet grew and evolved, it has gotten toxic, exactly like the places you would run away from before."[39]

The first generation of tech entrepreneurs was often comprised of those who didn't fit in. When they got into tech, it wasn't necessarily for money—the idea was still radically untested. Mainstream companies and financial institutions regarded it as a novelty, not a major investment opportunity. But once the internet exploded it brought in a new generation of investors and entrepreneurs, many drawn by the outsized returns. These weren't necessarily experimental outsiders, but they were still largely men.

There are a lot of ways to look at diversity, but the most helpful way I have found is called "diversity debt." In the same way that engineers can accrue "technical debt" when they push out sloppy code, or business owners can accrue "bookkeeping debt" when they procrastinate their financials until tax time, companies can also accrue diversity debt over their life cycle. The more people your company hires until you have a diverse team (meaning an array of genders, LGBTQ, socio-economic backgrounds, ethnicities, ages, able-bodiedness, etc.)—the more diversity debt your organization has accrued, and the greater

39 Some incredible founders shared their insights with me in, "12 Leading Investors Explain Why They're Funding Sextech," *Forbes*, March 2019.

danger it is to your long-term success. The tech community has accrued a blinding amount of diversity debt, and it's going to take a while to pay it down.

When tech platforms are built by people with homogenous lived experiences, it creates problems. It limits the appeal of the platform, and excludes audiences that can help it grow and evolve. Without a diverse team, without elevating voices and experiences unlike yours, you're operating in a bubble. You can't see what's going on outside until it's too late. Until the bubble pops.

Because sextech grew outside that bubble, it attracted an incredibly diverse set of ideas, drawn from a more diverse set of experiences—women, queer people, nonbinary people, and people of color. In terms of access and potential, sextech is much closer to the original founders of the internet than it is to the homogeneity of the second internet boom.

Sextech isn't just about sexual wellness—it's a way for tech to move past diversity debt and heal its deeper issues.

Every day, we hear more stories about the distrust of big tech platforms—congressional hearings on Twitter, the Cambridge Analytica scandal at Facebook, YouTube's arbitrary demonetization of users, toxic doxxing and harassment at Reddit, and deadly white nationalist galvanizing on 8chan. When scandals break, they can threaten the entire platform.

Silicon Valley is often criticized for its lack of diversity, but many of the problems it's faced are precisely because un-diverse, homogeneous teams were unable to see the problem before it exploded. Thousands of users can file complaints, but without strong, diverse leadership teams, few people will recognize the importance and have the power to act accordingly.

Tech should have been built for humans. Instead, investors celebrate the addictive nature of gaming systems. More consent education, more diverse founders, more Team Human startups that are helping humanity can save tech by helping make it more relevant in the world.

I don't blame men. We built a society without thinking of the ramifications of a very small, limited group of people. And look—women, queer people, POC are suffering—but men suffer as well. We all deserve better.

It's time to rebuild the internet.

{ CHAPTER 5 }

MOVING BEYOND THE ORGASM GAP

A few years ago, I met with a very well-known founder whose business sold to Amazon for close to a billion dollars. The company was a live-streaming platform that enables gamers to watch and interact with other gamers while they play video games.

When I talked about O.school, he referred to sexual wellness as a "niche" market multiple times. Laughing, I stopped him and reminded him that he founded a company to help gamers watch other people play video games. Our company, on the other hand, had every sexually active adult on the planet as a potential customer. He laughed, too—and made some introductions. For that, I'm grateful.

Over and over, I've learned there are really smart people who have no idea how big this opportunity is. How could someone so successful have no insight into problems that millions of people have? How could someone think of sex as "niche"? In this chapter, we'll look at the most common sexual problems that people face, where they try to find solutions, and why the major dominant platforms, like Amazon, are very unlikely to win this space.

PROBLEMS GALORE

Problems with sex are common, but rarely talked about. We don't talk about it with our doctors, our partners, or even friends. We're either ashamed to be sexual, or ashamed that we're having problems.

When someone comes along who is knowledgeable and unashamed, the floodgates open. Since founding O.school, I can't go to a dinner party or a meeting without being pulled aside and presented with a laundry list of issues keeping people from a satisfying, pleasurable sex life. Trouble at work. Having a baby. Menopause. Infidelity. A crisis in gender identity.

Keep in mind that some of these people I barely know. There's a desperate need to find a way to talk about these issues.

Consider this: Fifteen percent of marriages are considered "sexless," which means no sex for six months to a year.[40] I met an entrepreneur in the mental health space who shared that he had met a surprising number of men taking antidepressants like SSRIs to effectively "chemically castrate" themselves to keep sane—and faithful—in their sexless marriages. I was floored.

Unlike something like heartburn, the causes of sexual issues are varied and personal, ranging from ineffective partners to trauma to lack of personal knowledge to very specific physical issues.

Three out of four women have experienced pain during sex,[41] and for approximately 10 to 20 percent of women in the US, dyspareunia causes sex to be a regularly painful experience.[42] Vaginal dryness is a problem for 17 percent of women between the ages of eighteen and fifty.[43]

Men are struggling too—as many as one in three say they experience premature ejaculation.[44] According to the Cleveland Clinic, half of all men experience some form

40 Jen Gunter, "When the Cause of a Sexless Relationship Is—Surprise!—the Man," *New York Times*, March 2018.

41 American College of Obstetricians and Gynecologists, "When Sex Is Painful," September 2017.

42 Dean Seehusen et al, "Dyspareunia in Women," *American Family Physician,* October 2014.

43 Women's Health Concern Fact Sheet, "Vaginal Dryness," British Menopause Society, Accessed online June 2019.

44 WebMD, "Premature Ejaculation," Accessed online June 2019.

of erectile dysfunction at some point in their life, usually after the age of forty.[45] As with women, the causes can range from the psychological to environmental to high-blood pressure to drug and alcohol use.

If you have a sexual problem today, where do you go? If you need a vibrator today, where do you learn about it? If you are having issues with sex, what resources can you access?

CAN MEDICAL PROFESSIONALS HELP?

Carol Queen, the respected sexologist and co-founder of San Francisco's legendary Center for Sex and Culture, talks about the way the American public thinks about sexuality. She says that most often, we believe the most credible people to ask about sex will be our doctors. But, in fact, the average medical doctor gets fewer than ten hours of sex education during medical school.[46] Experts in mental health have told me that mental health professionals, therapists, and counselors also generally lack training to address gender and sexuality concerns. Meanwhile, experts like Carol have been largely ignored, regarded as someone obsessed

45 Paolo Capogrosso et al, "One Patient out of Four with Newly Diagnosed Erectile Dysfunction Is a Young Man—Worrisome Picture from the Everyday Clinical Practice," *The Journal of Sexual Medicine,* May 2013.

46 DS Solursh, et al, "The human sexuality education of physicians in North American medical schools." *Int J Impot Res.* 2003 Oct;15 Suppl 5:S41-5.

with sex rather than a true expert. In reality, she's got over two decades of research and cultural work behind her—and much less shame in talking about it than your doctor.

So many doctors over the years have not been able to clearly identify the anatomy of the clitoris and they often still debate the biological basis of the female orgasm.[47] Doctors just don't know enough about pleasure and sex to do more to help improve our experiences.

Most people don't feel comfortable talking to their doctor about sex and pleasure anyway. There's a perception that we're taking our questions to our doctors, but that's not really happening. While in one online survey, 54 percent of women reported that they would like to seek help from a physician for sexual function complaints, 40 percent reported that they did not seek help from their doctors.[48] And when we survey doctors, we find that these concerns are not routinely addressed.

In a comprehensive national survey of US obstetrician-gynecologists regarding communication with patients

47 Vincenzo Puppo and Giulia Puppo, "Anatomy of Sex: Revision of the New Anatomical Terms Used for the Clitoris and the Female Orgasm by Sexologists," Clinical Anatomy 28:293–304 (2015).

48 Laura Berman, et al, "Seeking help for sexual function complaints: what gynecologists need to know about the female patient's experience," Fertility and Sterility, Volume 79, Issue 3, Pages 572–576, March 2013.

about sex, only 40 percent routinely ask questions about sexual dysfunction, only 29 percent ask about sexual satisfaction, and only 14 percent asked about pleasure with sexual activity. For those who do ask questions, the doctor's miseducation may be detrimental.[49] People go in to say they are having painful sex, only to be told to drink some wine and relax.[50] This can be frustrating for straight women who feel misunderstood or shamed, and can be dangerous and life-threatening for people who are queer, trans, or sex workers.

A doctor shared a common scenario in the medical community with me—a straight woman goes to a physician and says, "When I have sex, I don't have an orgasm. My husband says I'm broken." The doctor finds there's nothing wrong physically, but determines there might be education or counseling needed. The patient can't actually afford a therapist, so the doctor suggests trying clitoral stimulation. While that could potentially treat the symptoms, the patient doesn't know how or what that means and neither the patient nor the doctor is particularly comfortable discussing it. Where can the doctor send them? The porn store down the street?

---

49   Janelle N. Sobecki et al, "What We Don't Talk about When We Don't Talk about Sex: Results of a National Survey of U.S. Obstetrician/Gynecologists," *Journal of Sexual Medicine*, Volume 9, Issue 5, May 2012.

50   Mikaela Conley, "What is Vulvodynia?" BBC Future, July 2018.

Definitely not.

Instead, the doctor might try to help by telling her to talk to her partner about it, but she isn't comfortable doing that either. After all, the doctor is as uneducated about it as she is. She leaves defeated and resigned.

There's a huge loop of inaccessibility, lack of information, and often layers of shaming that can make asking more harmful than doing nothing at all. There's a multiplying effect that happens when a trusted person in society rejects, shames, and misdirects.

So you're too nervous to ask questions, but your doctor doesn't ask you so you have to bring it up. When you do, the answers come from untrained, often shaming, perspectives. Layer on conservative politics, racial bias, gender issues, and the potential risks for queer people to be referred to "rehabilitative" therapy, and doctors and other medical professionals simply can't be our solution to sexual wellness. Not yet.

CLOSING THE ORGASM GAP

Everyone knows about the wage gap—it's a regular staple of political conversation. We don't talk nearly as much about the orgasm gap.

The orgasm gap is the discrepancy between men and women in how often they have an orgasm.[51] Every hundred times a straight man has sex, he'll orgasm around ninety-five times—about 95 percent of the time. Gay men and bisexual men reach orgasm between 89 and 88 percent of the time, and lesbians 86. But for a straight woman—a woman having sex with a man—the number drops to 65 percent (which according to some surveys, may still be too high).[52]

We've often been told that women have difficulty orgasming, or aren't as naturally sexual as men. As such, we don't prioritize women's sexual pleasure—instead, we blame the victim. But studies have shown that women don't have the same problem getting off when they're alone, or when they're with other women.

The orgasm gap is just another symptom of our miseducation from childhood thanks to taboo, mainstream porn, and bad pop culture scripts.

The overall message tends to be that men should initiate, that orgasm only happens through penetrative sex, and

---

51   Sadly, the studies did not cover more gender identities and sexual orientations beyond straight, gay, lesbian, and bisexual people who identify as men and women. More research is definitely needed in this arena.

52   Frederick et al, "Differences in Orgasm Frequency Among Gay, Lesbian, Bisexual, and Heterosexual Men and Women in a U.S. National Sample," *Arch Sex Behav.* 2018 Jan;47(1):273-288.

that sex ends when a man climaxes. We're not taught about arousal, lubrication, or the clitoris. We're not taught or modeled communication or consent. We're not taught to navigate things like painful sex or sexual activity after trauma. If porn is teaching our generation about sex, porn is not teaching us how to actually have sex that feels good or works for different bodies.

Pointing out the orgasm gap isn't meant to shame you if you're struggling with this. It's actually the opposite: you're not alone! Sex can be enjoyable without orgasm, and at the same time, your partner should prioritize your enjoyment, desire, and wellness. Statistics and experience tell us that instead of enjoyment, we're often faking orgasms and even experiencing discomfort with no path to resolution.

RETAIL EXPERIENCE

A large chunk of this industry—I'd say about half of the sextech purchases—happens offline in retail environments. To answer that call, there are a growing number of feminist sex stores like Good Vibrations and Babeland, and plenty of individually owned stores around the country.

A great retail experience will carry toys, lube, and all kinds of sextech innovations like Unbound, Dame, and Crave.

There are wearables, like Ohnut, that help with painful sex. There are apps, like B-wom, for pelvic floor health.

The majority of retail stores that sell sextech, especially in the middle of the country, are still stores like Hustler Hollywood and giant warehouses. They might provide the same types of sexual wellness products as something like Good Vibrations, but the experience is entirely different.

Typically, these kinds of stores will carry products with packaging that's covered in explicit images and marketing directed toward men. They can be pretty heavily misogynistic in their presentation, and all of it very much in your face. Some retail stores go beyond sexist products and unhelpful clerks. There might be places for people to watch porn and masturbate, or watch peep shows. Instead of feeling like a place for wellness, they often feel dark and shady and unsafe, especially if you're a woman.

If you have access to a feminist sex store, the experience is totally different. The clerks will be trained sex educators, which is incredible. As I described, I bought my first sexual wellness product at Good Vibrations in Berkeley, and I could see myself going there with a group of friends and having fun shopping with them. It was amazing.

Women have always preferred in-person education when it's available. In Vibrator Nation, Lynn Comella talks about this problem—before feminist sex stores, she writes, women just didn't go to them at all.

But well-lit, resource-rich stores like Good Vibes are located in a small handful of cities—inaccessible to most of the population. Rural and suburban communities often have zoning regulations that push any sex-related retail far outside the city center, often in an industrial section of town, or off-the-highway near a truck stop.

Thanks to the internet, people who already know what they want can go online to respected retailers like Adam & Eve, Good Vibes, or LoveHoney in the UK. Even Amazon sells sex toys—if you look closely enough. But buying online doesn't help people who want education, or who need more information before making a purchase.

WHY AMAZON WON'T WIN SEXTECH

Certain things are completely transactional to buy. Pretty much anything in your house—your clothes, makeup, shoes, food, charger cords, or kitchen supplies. You read product reviews, find an item, buy it, and that's it. Amazon has done an amazing job delivering on that experience. Outside of those products, there are some experiences where Amazon has a harder time "winning" (capturing) the market.

With an everyday transactional product, you can build the

right product, launch it, and go. People can hear about it, search for it, and buy it online. So why does Amazon have over sixty thousand individual products (known as SKUs) in sexual wellness categories, but only a billion dollars of the current $27 billion available in the market? Considering their dominance in other spaces, it's telling. This space of sextech and sexual wellness hasn't filled up yet because mainstream retailers don't know how—or can't—approach it.

Sexual wellness isn't alone in this space. You obviously wouldn't go to Amazon to buy insurance, for example, or the financial software we sold at my previous company, inDinero. Some products require more than a few Amazon reviews to help us feel confident in our choice. Amazon isn't great for products that require a lot of education and consultation.

Even beyond that, few people—especially those first starting out—have any idea what to buy. There are dozens of options for toys—how do you know what to trust? How do you know what will work best for your body? Reviews aren't enough—and how can you trust the suppliers? Many sex toy reviewers and writers have covered the problems with fake or counterfeit products sold on Amazon, and many advise against going to Amazon due to safety and quality concerns.

Many sex toy retailers have told me stories about people

purchasing faulty or counterfeit items on Amazon and trying to get help from the retail stores, only to be told they should buy from trusted suppliers to ensure they can take advantage of product warranties.

Not only are there many fake reviews, fraudulent and counterfeit products, and offensive marketing used to sell many sexual products on Amazon, the customer reviews aren't helpful either. Customer reviews are most helpful when customers know what they need, what they are looking for, and have experience with the products they are buying. As we covered in chapter 2, most first-time consumers may not know they need to check product warranties, what body-safe materials they are working with, what lubricants pair with what toys, and what toys will enhance the customer's desired activities the best. Since everyone's individual experience is so unique, customer reviews can be misleading, unhelpful tools to choose a product. And the wrong decision can often result in an expensive purchase that is, understandably, unable to be returned or exchanged easily.

In addition, for sexual wellness, it's important to help ensure that purchases can arrive in discreet packaging. There are many horror stories online about poorly wrapped phallic products that scream "I am a dildo" to the world. In an ideal world, this would be regarded as mundane as seeing a sleeping bag or yoga mat or other product on someone's doorstep. Today, this is still a huge concern and a barrier to purchasing toys online. Many modern retailers for sexual wellness products understand this and soothe people at checkout about their discreet packaging.

Not only that, but no one wants to be "retargeted" for dildos. Retargeting is the freaky phenomenon of seeing a product you may have viewed on another site show up in the ad section of another completely unrelated site. Retargeting is one of the ways that Amazon has been so successful. But retargeting is one thing for cookware or a garden hose—and quite another for vibrators or lube. What happens when you're browsing on your laptop at work or with your mom sitting beside you?

A few years ago, a man walked into Target furious. His teenage daughter was receiving a flood of coupons for cribs and diapers—was the company trying to encourage her to get pregnant? It turned out that, unbeknownst to her father, the young woman was pregnant, and had been purchasing products online related to early stage pregnancy. Retargeting had outed her.

Imagine if the company had done the same with a sexual wellness product like lube or a vibrator. Companies like Amazon depend heavily on retargeting to increase sales, but when it comes to sex-related products, they're effectively silenced. They won't even list them on the home page. They won't always come up during search.

Shame and fear won't go away easily enough to fix this problem. We need to fix the marketing. We need to fix the delivery. And despite the size of the market, existing large-scale companies are neither incentivized nor prepared to go after it.

These factors keep Amazon from totally dominating the market. If we had a shame-free sex educational system, if we had medical doctors that could talk about pleasure in a knowledgeable way, if we could have discussions with our partners without embarrassment, if we could ask friends for recommendations, the industry could just focus on making the best products and then put them out there in the world to be purchased. As it is, we all have to do some education and de-stigmatization no matter what part of the sexual wellness market you're working in. People who have been raised in a conservative way, which is a huge part of the population, have to go through a process before they will even be open to buying a vibrator.

The company that will win sexual wellness isn't a current dominant player. It won't be Amazon or any of CPG brands we know today. It will be a new entrant.

It's going to be one of us.

{ CHAPTER 6 }

THE NEXT GENERATION

This year, I had dinner with some pioneering founders in reproductive health, many of whom I deeply respect and consider brilliant. But when I mentioned identifying as queer, these predominantly older, straight white women—women who'd been on the leading edge of women's health for much of their careers—asked me, "Why do you say queer? Isn't that a bad word?"

They asked me why I introduced myself with my pronouns, why I preferred "reproductive health" over "women's health," what "non-binary" was? They even asked why I preferred menstrual cups—weren't they gross? Sometimes they used the "wrong" words. It was eye-opening. Here I was, surrounded by pioneering women, having conversations eerily reminiscent of the conversations I've had with all-male investors, "Why do you use the term vulva and not vagina?"

I had empathy for them.

While they were busy fighting to even say the word "vagina" and delivering innovative health products to underserved communities that they knew well, the world changed. Awareness about gender and sexuality has shifted in the last decade—and fast.

Yet again, I was deeply reminded of why I started O.school—because even as a millennial who went to a liberal university in Berkeley, California, my understanding of gender and sexuality was also still limited. How could I blame them? I was once looking for answers too, and in many ways I still am. "Wokeness" is a lifelong journey. Every day I learn about ways to evolve my language, practices, and values.

When I started out, O.school was a resource I desperately needed for myself. O.school's official incorporation birthday was the day before I left for my last 500 Startups business trip, a week in Kobe, Japan. It was my last hurrah before taking the full-time plunge into the startup trenches again. On my first day in Japan, I met a charming, quiet Chinese person, and I fell madly in love. Before I started working in tech startups, my plan was to move to China. I studied abroad in China and, with this person, my Mandarin came alive again. Using bits of my native English and their native Mandarin, we got to know each other.

Their gender identity felt completely ambiguous to me, so I remember coming out to them first, "I'm queer." We were on a date, beaming, eating Chinese food in a small town in Japan.

In as much English as they could, they said, "I am not lesbian."

Laughing and trying to communicate in my rusty Mandarin, I said, "Everyone is on the menu for me, I like everything." They laughed, relaxed, but still struggled to find words.

"Are you trans? Non-binary? Genderqueer?" I asked. We had to pull out Google Translate on our phones to bridge the gap.

I'll never forget that moment. They looked up from the translation app and said, "Yes. Trans! You're the first person who's ever understood. My family still calls me a lesbian. But that's not what I am!" Language had been a barrier, but we saw each other in that moment. Being seen and loved for who you are, creating a safe space for someone you love, is a powerful, healing experience.

However, I had never had an out trans partner at that time. Later, alone, I tried Googling for help about what to do and not to do with a trans lover. I found very little.

I had no idea how to take care of this person, and we had limited communication due to our language barriers. I returned to the United States humbled, understanding how much more I had to learn. And that I needed to create a space where others could grow and change, too.

When the women at dinner asked me why I identify as queer, it's moments like these that came to mind—when gay didn't fit and I needed other words to describe who I am and whom I'm open to loving. I was happy to be a safe space for them to ask questions that day. One day, I hope I'll be able to talk to them about how "women's health" may be a term we swap in favor of a term more inclusive like "reproductive health," but it wasn't the right dinner to bring that up just yet. Learning this language was a process for me, and I understand that it takes time. I hope to be a safe resource for people to learn about the changing language and values around gender and sexuality.

This chapter is for all innovators, especially sextech businesses, who want to prepare for the changing world.

OUR CHANGING DEMOGRAPHICS

When you remove shame from the equation, a very different picture comes into focus and the business opportunity becomes clear. While we face many challenges, demographically we're moving toward a more open mindset as a society.

As Rebecca Traister wrote extensively about in *All the Single Ladies*, for the very first time in American history in 2009, unmarried women outnumbered married women.[53]

People are staying single longer, having more sexual partners, and using more toys both with and without partners.[54]

Millennials are twice as likely to identify as LGBTQ as any other generation, with Gen Zs trending even higher.[55]

Today's generation is more likely to be single and have less sex. Younger people, Gen Zs, are waiting longer than ever to have sex.[56]

Then there's the question of gender, which shakes up

53 Rebecca Traister on Fresh Air, "Single By Choice: Why Fewer American Women Are Married Than Ever Before," NPR, March 2016.

54 US Census Bureau, "Median age at first marriage: 1890 to present," November 2018. Nicholas Wolfinger, "Nine Decades of Promiscuity," Institute for Family Studies, February 2018.

55 GLAAD, "New GLAAD study reveals twenty percent of millennials identify as LGBTQ," March 2017.

56 Youth Risk Behavior Survey, "Trends in the Prevalence of Sexual Behaviors and HIV Testing National YRBS: 1991—2015," CDC, 2015.

everything the current industry knows about marketing sexuality:

- One-third of GenZers know someone who uses gender-neutral pronouns.[57] And according to one study, there are now 700,000 out trans people in the United States.[58]
- About one percent of the population identifies as asexual,[59] and another 1.7 percent of the population is intersex.[60]
- Non-traditional relationship structures like poly-amory are on the rise with over 12 million individuals reporting trying a non-monogamous relationship structure, about 4-5 percent of the heterosexual adult US population.[61]
- And, very important for sextech, half of American

---

57 Kim Parker, Nikki Graf, and Ruth Igielnik, "Generation Z Looks a Lot Like Millennials on Key Social and Political Issues," PEW Research, January 2019. Shamard Charles, MD, "Teens Are Having Sex Later, Using Contraception, CDC Finds," NBC News, June 2017. Joyce Abma and Gladys Martinez, "Sexual Activity and Contraceptive Use Among Teenagers in the United States, 2011-2015," CDC, June 2017.

58 The Williams Institute, "What We Know: Transgender Population Research Guide," UCLA School of Law, June 2015.

59 LGBTQ Life at Williams, "10 Things you need to know about Asexuality," Williams College, Accessed online June 2019.

60 Hida Viloria, "How Common is Intersex? An Explanation of the Stats," Intersex Campaign for Equality, November 2017. L Sax, "How common is intersex? a response to Anne Fausto-Sterling," *J Sex Res.* 2002 Aug;39(3):174-8.

61 Ethan Czuy Levine et al, "Open Relationships, Nonconsensual Nonmonogamy, and Monogamy Among U.S. Adults: Findings from the 2012 National Survey of Sexual Health and Behavior," *Arch Sex Behav.* 2018 Jul; 47(5): 1439-1450.

women have tried a sex toy, and this number is rising.[62]

To have a healthy society, people need to be able to express their sexuality and gender without being shamed or judged. Even what may seem minor to the mainstream world can have a huge effect. When young transgender people in all parts of the country are able to use their chosen names, they experience 71 percent fewer symptoms of severe depression. 34 percent report a decrease in thoughts of suicide. There is a 65 percent decrease in suicide attempts.[63] Affirming gender identity by using chosen names in multiple contexts allows kids to settle into who they are and improves their mental health overall.[64]

Even when coming out is stressful, the long-term effects are positive.[65] It's only stigma and discrimination that create a mental health struggle for LGBTQ youth.[66]

---

62 Debra Herbenick, "Prevalence and Characteristics of Vibrator Use by Women in the United States: Results from a Nationally Representative Study," *Journal of Sexual Medicine*, July 2009.

63 University of Texas, "Using Chosen Names Reduces Odds of Depression and Suicide in Transgender Youths," *UT News*, March 2018.

64 Stephen Russell et al, "Chosen Name Use Is Linked to Reduced Depressive Symptoms, Suicidal Ideation, and Suicidal Behavior Among Transgender Youth," *Journal of Adolescent Health*, October 2018.

65 Stephen Russell and Jessica Fish, "Mental Health in Lesbian, Gay, Bisexual, and Transgender (LGBT) Youth," *Annu Rev Clin Psychol.* 2016 Mar 28; 12: 465-487.

66 Human Rights Campaign Foundation, "Mental Health and the LGBTQ Community: LGBTQ Youth and Mental Health," Suicide Prevention Lifeline, Accessed online June 2019.

In short, the world is changing, and more awareness about LGBTQ issues is good for public health and good for business. Businesses, entrepreneurs, and innovators who want to build for future generations should pay attention to changing demographic trends.

THE FUTURE IS QUEER

Gender is changing, and much of the world is still struggling to catch up. In the last ten years, the entire conversation about gender has shifted. Every year, I'm invited to speak at the first-year orientation of colleges all over the United States. Every year, I see our teens changing more and more.

Recently, after one college event, someone was telling a group about the gender reveal party for their sister's upcoming child. One of the kids, who was seventeen, was honestly perplexed and asked, "How do you know the baby's gender? Don't you mean the sex?"

Many of the Gen Z audiences that I have met question gender like this, thinking more deeply, critically, and thoughtfully about the binaries and limitations we've been prescribed. Meanwhile, in other parts of the country, conservatives are still talking about LGBTQ people like they do not exist. There are actually seven states in the US where teachers are prevented, under penalty of law, from acknowledging LGBTQ people in school outside of the context of AIDS or to condemn homosexuality.[67]

Traditionally, people start to form ideas about who they're going to be based on the sex they're assigned at birth. It's such an ingrained norm that it affects the way spaces are organized and designed, the way people communicate, and the assumptions people make. Pushing back on gender is one of the most fundamental ways to push back against society, and there's massive fear around what that means. I don't have any answers about what it's going to look like for you—if you're not trans or non-binary or genderqueer or you do not know anyone who is, you may not consider this change crucial. But it is. And if you're working in sextech or sexual wellness, you'd better have people with those lived experiences represented in your company.

To become a ten-, twenty-, thirty-year brand, companies have to be prepared to serve millennials and Gen Z markets.

Big brands already see awareness around gender shifting and are trying to adapt. Like the infamous Gillette toxic masculinity ad—in which the razor company controversially tried to address issues like bullying and sexual

---

67   Research Brief, "Laws that Prohibit the "Promotion of Homosexuality"": Impacts and Implications," GLSEN, Accessed online, August 2019.

harassment, and received vicious conservative backlash. (Unbowed, Gillette then released an ad with a father teaching his trans son how to shave.)

It's why makeup brands have started including male, nonbinary, and trans influencers in their campaigns. Reactionaries derisively consider this virtue-signaling, but huge global consumer product brands like Proctor and Gamble, the massive $66B+ revenue company that owns Gillette, Pampers, and many brands we all use every day, are more cynical than virtuous. They see the coming change in their markets.

Millennials and younger people have either embraced non-binary identities or know someone who has. They have friends and family who are trans, or non-binary, or LGBTQ, or they are themselves. Bigger and bigger markets are growing to support businesses who support the people in these younger lives. Without growing alongside our audience, we'll lose them.

A SEX RECESSION?

Frequency of sex is not an indicator necessarily of society's sexual wellness, but in the past few years, something interesting is happening.

In 2010, The Kinsey Institute found that eighteen and

twenty-nine-year-olds have on average 112 sex sessions per year, or twice a week. Thirty to thirty-nine-year-olds have sex 86 times per year, which equates to 1.6 times per week. Forty to forty-nine-year-olds have sex 69 times per year, about half the total for eighteen to twenty-nine-year-olds.[68] The same source indicates that married people have more sex than single people (this surprises some), and single people masturbate more than married people.

Yet, recent studies have reported that younger people are having sex less than previous generations did at the same age. Between 1991 and 2017, the number of high schoolers who had had intercourse decreased from 54 percent to 40 percent, and another study also showed that people in their twenties were 2.5 times more likely to be abstinent than previous generations of the same age.[69] The Atlantic recently—and controversially—called it a "sex recession."[70] Young people are doing the opposite of what people would expect in a world with more sexual openness and the ubiquitous rise of hookup apps like Tinder.

68 Lizette Borreli, "Am I 'Normal?' Average Sex Frequency Per Week Linked To Age," Medical Daily, August 2017.

69 Youth Risk Behavior Survey, "Trends in the Prevalence of Sexual Behaviors and HIV Testing National YRBS: 1991—2015," CDC, 2015.

70 Kate Julian, "Why Are Young People Having So Little Sex?," The Atlantic, December 2018.

There are many potential reasons for this, starting with the battle for our free time that previous generations didn't experience. We have access to more social media, more porn, and more gaming, all taking up our time and focus. Then, there's the new relationships to technology. On a sextech panel recently, Myisha Battle, a certified sex and dating coach and host of the *Down for Whatever* podcast, told me and our audience about a new modern barrier to intimacy: mobile phones. Her recommendation to keep devices out of the bedroom garnered a loud groan from the multi-generational audience.

Technology is definitely one aspect of this. "Digisexuality" is a term that Myisha introduced me to, describing the emerging research about human sexual connection and technology, and the identical brain chemicals that some people are getting from interactions with technology, rather than human relationships.[71] That's one popular angle about the sex recession, that our generations will end up like Japanese youth who have reportedly turned to digital romantic partners.

But, like Japan, the economy and cultural shifts are also large drivers. More young people are living at home with their parents, and even if they aren't, there's more economic stress all around—even in a theoretically strong

---

71   Neil McArthur, "The rise of digisexuality: therapeutic challenges and possibilities," *Journal Sexual and Relationship Therapy*, Volume 32, 2017.

economy, wages haven't kept up with inflation, let alone home prices or healthcare.

We're simply more preoccupied as a generation.

When I'm asked about this, however, people are often shocked to hear that I don't worry about how much sex people have. What's more important to me is if their experiences are positive and pleasurable.

Just as old employment metrics have struggled to handle the rise of the gig economy, I wonder whether these studies are acknowledging the queerness of modern generations, or non-traditional sexual activity, like manual sex with hands, masturbation, mutual masturbation, or other forms of physical connection. In the Atlantic piece, it was stated that oral sex was considered in the study, but the sexual menu has expanded in recent decades to more than just intercourse and oral sex. It would make sense to me, knowing that over 70 percent of vulva owners prefer or require clitoral stimulation to achieve orgasm, that less traditional intercourse may signal that other clitoris-focused activities were happening, which would be one positive step toward closing the orgasm gap.

People are also becoming more educated about asexuality, BDSM, or kink (which may or may not include intercourse), and intimacy-building activities like cuddle parties gaining in popularity. People of all generations, but especially Gen Z, are redefining what it means to be sexual.

While the Atlantic asked "Why Are Young People Having So Little Sex?" and covered the sex recession, the BBC recently asked, "Are We Set For A New Sexual Revolution?" in which David Halperin asked, "What is sex for?" and inquired about the new purpose of sex. Especially with birth control, the growing popularity of egg freezing and in vitro fertilization, and decrease of sexual taboos, what if the "sex recession" is one of the signals of the next sexual revolution?[72] This generation is opening the discussion: what constitutes sex?

At a college event last year, I received an anonymous question that I'll never forget. It went something like this:

"I am a sophomore, and I have been sleeping in the same bed as someone on my floor for a few months. We cuddle, but don't have sex. Is this normal? Should I stop it? How would I stop it?"

They wanted intimacy, but not sex—and felt bad about it. There are many reasons why people may be having less intercourse, but I do know they're having different

---

72   Brandon Ambrosino, "Are We Set for a New Sexual Revolution," BBC Future, July 2019.

sex, and we need to catch up in our studies and analysis to be more inclusive and relevant in our modern world.

NAVIGATING CULTURE WITHOUT FEAR

The world which builds products for women and products for men is going to change. As much as it makes people nervous, we're not living in a binary world anymore.

Building a sexuality brand, which is so intrinsically tied to gender in all its forms, can be scary. I can see the perspective of the previous generation of "women's health" companies struggling to make sense of it—they worked so hard to make sure vaginas and vulvas got respect, and here we millennials and Gen Z folks are skipping right over them and past the binary. Insisting we refer to vulva owners instead of women, and understanding that any gender can have a vulva or clitoris.

Millennials are not wrong in asking for these things, but we couldn't do it without the work of those who came before. That's a big jump, and some people still need time to heal.

We need to communicate that this is a "yes, and" issue, where yes, we need to keep talking about breasts and pussies and do it in a way that doesn't invalidate trans and non-binary people. Janelle Monae's PYNK music video is a great example. The video is dedicated to celebrating all things "pussy"—with trans women invited to the celebration. That's what we need: to celebrate each other, hold space for each other, and fight battles together.

The impact of these changes on society are not lost on me. There is much structural work to be done to adapt government, business, and society to the changing demographics. Massive software overhauls, database overhauls, and huge systems are based on gender markers (which is how gender is labeled on an ID)—from Google Analytics to Customer Relationship Management software like Salesforce to the way all insurance is underwritten—all have massive databases with entries that are categorized based on gender. Medical research will have to evolve. Moving away from just "man" and "woman" to a world where people can be represented is going to be a process. We see states changing their gender markers, and we will continue to see more.

At O.school, we started by degendering our site architecture, organizing our content by body part, rather than gender. That early DNA of our site has been something we've carried through—the most recent example? Our digital "orgasm order forms"—notes you can send a prospective partner detailing what you like or don't like—is based on whether you have a penis or a vulva, rather than treating them as synonymous with a specific gender. As a sextech entrepreneur, your product or content or wellness focus might find a different approach. We're well aware that our own approach could, in time, seem outdated.

Some businesses and platforms are there to bridge the gap between older conservative generations and this one. O.school is here for the people who don't have a gender studies degree and have never met a non-binary person. We are here as a safer space for them as they work through these issues, as much as we are here for the underrepresented.

MOVING FORWARD AFTER #METOO

In 2018, Tarana Burke's #MeToo movement sparked a massive reckoning in society. No one was off limits. Weinstein, Cosby, Louis C.K., Hollywood, academia, tech, politicians, college campuses. The President of the United States, as of the writing of this book, has at least twenty-four allegations of sexual assault against him.[73] There are a host of other hashtags, such as #WhyIDidn'tReport and #TimesUp to help propel the conversation. We're getting clear about the fact that sexual misconduct, ranging from sexual harassment to sexual violence, pervades every level of our culture.

---

73   Eliza Relman, "The 24 women who have accused Trump of sexual misconduct," *Business Insider*, June 2019.

It's about time.

One-third of all women are going to have at least one traumatic sexual experience at some point in their lives.[74] Half of all trans people are sexually abused or assaulted in their lifetimes.[75] One in four men have experienced physical sexual abuse at some point in their lives.[76]

The #MeToo movement began a crucial conversation, unleashing pain and rage that had been suppressed and ignored for too long, especially but not only for women.

Sexual wellness advocates need to educate themselves about how to provide trauma-informed spaces, experts, and resources for people who have experienced trauma. Sadly, it's a big demographic. This can include, but is not limited to, consulting with trauma-informed care experts, content or "trigger" warnings, and learning about trauma-informed consent guidelines for medical professionals, educators, organizations, and businesses.

A big part of the future of sexual wellness is supporting healing for those who have experienced sexual trauma.

---

74   National Sexual Violence Resource Center, "Get Statistics," Accessed online July 2019.

75   Office for Victims of Crime, "Responding to Transgender Victims of Sexual Assault," Office of Justice Programs, June 2014.

76   The 1in6 nonprofit, "*The 1 in 6 Statistic,*" Accessed online July 2019.

And yet, despite the crisis, men seemed surprised, confused, even angry about the reckoning. In many cases, they didn't understand what was happening, or why. For those of us with a lived experience of harassment, abuse, assault, or other trauma, the movement made entire sense. But for many men—it was as if you'd told them that the sky was actually red. It didn't match their experience, or what they'd been taught.

Society has failed to educate perpetrators of sexual assault, especially men. Watch a Hollywood movie, check out most middle school or high school sex education classes, visit most college fraternities, and you will seldom find a strong foundation of consent education. Society teaches men to be sexually aggressive, to take initiative and make the first move, but not how to read the body language of others, make others feel comfortable saying no, or how to navigate power dynamics in professional and dating environments. If we want to reduce and eliminate sexual assault, we must teach everyone about the critical—and sometimes complex—elements of consent.

Men may know that the rules have changed, but that doesn't mean they know what the new rules are. A recent survey showed that 60 percent of male managers were uncomfortable mentoring women in the workplace, a 32 percent increase from the year before.[77] Many men that I have spoken to report feeling scared, anxious, and confused about how to be men in 2019.

States are willing to chemically castrate sex offenders, but won't invest in even basic sex education, let alone consent education. Politicians who ban abortion are equally willing to block their schools from teaching about sex and bodily autonomy. There are basic, simple, proactive things that we can do as a society to begin harm reduction.

We cannot banish men, or sexists—as appealing as it might seem—even all sex offenders. Even if we wanted to, it's not feasible. If we're to move forward, we need to find a way to build systems that educate and protect.

The sexual wellness industry must be part of that movement to bring education, healing, and support to shape a better future for ourselves and the next generation. As Frederick Douglass said, it's easier to build strong children than to repair broken men.

---

77 LeanIn.org, "Working Relationships in the #MeToo Era," Key Findings. Accessed online, August 2019.

CONCLUSION

You walk into Walgreens today and you see vibrating cock rings in the condom aisle. Netflix has a hit show called *Sex Education* about an awkward high schooler and his sex therapist mother, and Disney is beginning to introduce the next generation to more queer characters. The New York Times regularly covers open relationships. *Fifty Shades of Grey* had massive, mainstream appeal. There are still a lot of barriers to healthy sexuality in our society, but you've probably noticed the shift toward sexual wellness happening. The world is changing.

Sexual wellness is not about sex robots. It's not just about more pharmaceutical drugs. It's not about having more sex, less sex, kinkier sex, or any specific type of sex. It's about bringing us back to a holistic sense of wellness. It's about healing a lot of shame and trauma. It's about empowering people to live in their full power.

We have a nation that's not educated about sex. We have a nation where one-third of its people have experienced sexual trauma. We have billions of people running around the world who've never been taught about consent. Not only do we have to grapple with making the existing world safer for all people—to end the immediate threats of sexual violence globally—but we also have to shift our thinking so that these things aren't so hardwired into our culture anymore.

In the back of this book, I've included a section for budding sexual wellness innovators and sextech entrepreneurs. We'll explore the blue ocean and talk about how to navigate building a sextech company.

THE FUTURE OF SEXUAL WELLNESS

Imagine for a moment a future where we rely less on moral and religious institutions to set societal standards around sexuality. Instead, we rely on public health institutions and evidence-based standards to determine what is taught in schools and the required training for medical and therapeutic professions. Doctors and teachers are trained to serve diverse needs. We shift away from puritanical, harmful, shameful views on sexuality. Education evolves. Kids are taught comprehensive sex education including consent, masturbation, and medically accurate anatomy. Sexual wellness products, including vibrators and lube, are prescribed by doctors. The health and wellness industry includes physical health, mental health, and sexual health.

Imagine a queer, nonbinary, and trans-friendly future. "Coming out" is no longer a thing. Doctors, nurses, and therapists are trained to help people of all ages with gender-related care. Gender nonconforming spaces are the norm. They/them pronouns are widespread. Pop culture has a rich representation of many types of genders, sexualities, and identities. Porn is better. Sex work is respected, decriminalized, and safer. A new masculinity emerges. Men are allowed to be sensitive and form healthier relationships to gender and sexuality. Sexual assault rates decrease. The orgasm gap is in the past.

People have access to their power through education, self-acceptance, and community.

This is the future of sexual wellness—if we build it.

BONUS GUIDE

NAVIGATING THE BLUE OCEAN OF
SEXTECH AS AN ENTREPRENEUR

"I didn't start a business to sue anyone," says Alex Fine, a fellow sextech founder and friend, and I feel for her.

Alex's company, Dame Products, an innovative sexual wellness toy company, filed a lawsuit against the New York City's Metropolitan Transit Authority (MTA) for discriminatory advertising policies in June, 2019.[78] The MTA denied Dame's tasteful ads on the basis that they violated a ban on ads for "sexually-oriented businesses"—even as they approved a massive ad buy for Hims, which provides erectile dysfunction medication. Anyone who's ridden

---

78 Estrella Jaramillo, "Sextech Company Sues NYC's Subway for Discriminatory Ad Censorship," *Forbes*, June 2019. Sara Ashley O'Brian, "Startup that makes sex toys for women sues New York transit system for banning its ads," CNN Business, June 2019.

a New York Subway recently is likely familiar with the not-so-subtle erect cactus ads. Or ads for Viagra, breast augmentation clinics, or the Museum of Sex.

Dame's lawsuit is a stand against advertising bias against sextech, and the issue Alex faces as a founder is far from unique.

Unbound, another female-founded sextech company found itself in a similar bind with the MTA, eventually suing and winning. (The ads floundered, however, as the MTA refused to agree to a design.) Thinx, a femtech company that makes menstrual underwear, and many other companies, have spoken out about the difficulties advertising products that have anything to do with the vulva.[79]

At O.school, we've had our issues, beyond Facebook bans or Instagram flagging. A few years ago, we noticed that our own site was being blocked by Apple's new parental controls, intended to keep kids away from porn. O.school and gay teen suicide hotlines were blocked, but semi-explicit Maxim.com, and Nazi sites like Daily Stormer, with articles advocating the rape of feminists, were not.[80] The double standards are endless, and every sexual wellness company has countless similar stories.

---

79 Salvador Rodriguez, "Facebook is blocking ads that target women with menopause but allows ads from companies selling pills for erectile dysfunction," CNBC, March 2019.

80 O.school, "Censorship And Sex Ed," 2019.

It's part of the reason we're all so close.

When sextech founders get together, the most common topic of discussion is not how fun it is to play with sex toys all day, or our passion for sexual wellness, or the thousands of people we have helped—it's how we are all navigating the discriminatory advertising policies of platforms who disapprove and ban even our most carefully crafted, modest, rule-abiding ads.

Unlike many startup companies, we know we are building things that people want. We know how much our customer bases love our products or need our content. The struggle is getting the message out and reaching those audiences. Advertising challenges are a big deal for any company—if you cannot acquire new customers sustainably, it's harder to predictably grow your audience month over month, and thus, raise additional growth capital.

These restrictions are one of the reasons the bigger players haven't moved into the space. The terrain has historically been tough—which, conversely, means there's massive opportunity.

In entrepreneurship, a blue ocean opportunity means that there aren't many competitors. The possibilities are as limitless as the horizon. (The opposite would be a "red ocean," because of all the blood in the water from the battling.) The "blue ocean" market of sexual wellness takes guts, fortitude, creativity, and patience.

For sexual wellness entrepreneurs, there are many types of business models. You can be an influencer on Instagram who creates educational content and makes money from speaking opportunities and affiliate links. You can build a retail store to help serve people's needs. You could work in advocacy and public policy. Or, you can build a startup company in the private sector.

As a sextech entrepreneur, I'm thrilled to support those in every aspect of sexual wellness, from rural LGBTQ activists to healthtech innovations like pelvic floor products to at-home STI testing kits to artisan glass dildo makers. Our paths may all be different—and each path has unique advantages and disadvantages—but our goal is the same. As a startup founder, I depend on all of these to further open the conversation and destigmatize sex.

But if you're like me and want to build a tech or startup venture that takes on this blue ocean, follow along and I'll tell you what I know.

HOW TO RAISE CAPITAL

Raising capital is very important for a blue ocean company, because it often takes more time to develop a business model. Raising money is difficult for most startups, no matter what they do—raising in a blue ocean market raises the stakes.

People used to only talk about how difficult raising money is for sextech, but in recent years, as more companies lock down larger numbers, that's begun to shift. Sexual wellness in particular has seen more investment than ever.

In February 2019, Dipsea, a short and sexy audio stories company, raised $5.5M in the massive—but long unrecognized—audio erotica space.[81] More and more investors are looking at sexual wellness.[82] Unbound, Dame Products, Maude, Crave, Lioness, Bloomi, B-wom, Lora DiCarlo, have all raised rounds, according to Crunchbase.

The momentum is there, as is the opportunity. But with any blue ocean startup, the real issue is how to build the boat.

Building a talented team requires upfront capital, unless you can get the best people to work only for equity—ownership in the company, usually through stock—which is

---

81   Kate Clark, "Dipsea raises $5.5M for short-form, sexy audio stories," *TechCrunch*, February 2019.

82   Twelve of them talked with me in, "12 Leading Investors Explain Why They're Funding Sextech," *Forbes*, March 2019.

very difficult. Office space, legal services, product development, travel—these all cost money.

So, how do you fundraise? There are many resources devoted specifically to this, but I will give you a short playbook here.

First, the best advice before launching a startup is to spend a solid month convincing yourself why you should not start that business. Research all of your competitors. Tell yourself all the reasons you won't succeed. Sit with all of the ways that you'll become one of the many startups who will inevitably fail. Tell yourself these things, because it's likely to be the truth. And, like it or not, investors, potential hires, and pretty much everyone you know are going to want to know the answer as to why you think you might be different.

Get therapy if you can afford it. Become self-aware. Make sure you really, really, really want to do it. Then, do a market deep dive. Try everything. Talk to potential customers about your idea. Talk to competitors. It's not all or nothing, remember—if founding your own startup isn't right for you or the timing isn't right quite yet, you might consider working for the best founder you can find in your space, which, aside from being an excellent strategy to build a network for your future startup, is a great way to learn while still earning a salary. It is not necessary to go into personal debt for a venture, but it is very common. With my family background and lack of personal savings, it wasn't feasible to start full-time until I raised money.

O.school began as a side hustle passion project while I was still working in venture capital. During this time, I developed the concept, conducted market research, recruited people to help me, and secured funding before I quit and started working on O.school full-time.

Running a startup is really hard. You face constant failure, uncertainty, lack of resources, pressure, financial troubles. Make sure you're in a strong place emotionally, and ideally financially, when you start. Remember, it's going to get worse before it gets better—if it gets better.

You have to be okay with not seeing your scaled vision right away. You have to be okay with things rarely working or being enough. You have to be okay with things being broken and under-resourced and with constantly selling things or bringing things to market that aren't quite perfect. Keep the vision of what it can be in front of you, because it can be difficult to go forward when that vision isn't yet real.

There is no single playbook for fundraising, and the process will vary for every founder, business model, and stage. There are many existing comprehensive resources online from accelerators like YCombinator and 500 Startups, as well as venture capitalists who write, like Elizabeth Yin, Hunter Walk, Both Sides of the Table, and many more.

Reach out to other founders or investors you trust to talk about your fundraising strategy, and be prepared to discuss questions like:

- What is the size of your market? What's your target audience?
- What unfair advantage does your team have that other teams don't have?
- What does a minimal viable product look like for the problem you are trying to solve?
- Can you clearly describe the problem you will solve, and how you will know your solution addresses it?
- What are the key metrics for your business?
- What measures will indicate whether the business is working or not working?
- How much money do you need to raise right now, and why?

Let's talk about the last one, because it's probably one of the most critical questions.

Think of rounds of funding like gas stations. Putting more money into your startup is like filling up a gas tank in your car. First and foremost, you have to understand your "burn" and "runway." Understanding the "burn" of your company (the amount you will spend every month, after any revenue, if any) in conjunction with the amount of money you plan to raise will give you a sense of your "runway" (the number of months your company can operate until the next funding round). A company that raises $1 million dollars and spends $50,000 every month has twenty months of runway. A startup that is not profitable will likely die or need to think about exit strategies once their runway dwindles away.

You wouldn't want to fill up only enough gas to get you 20 miles down the road when the next nearest gas station is 100 miles away. No one wants to run out of gas in the middle of nowhere. It's the same with fundraising, since you do not want to raise an arbitrary amount of money without thinking about the next round of funding.

So, how do you know how much gas you will need in your tank, or how much runway you need?

You need to deeply understand the milestones required to raise the round after the current round you are trying to raise, and estimate how long it will take.

It sounds easy, but actually it is a step that requires a lot of research. It's also tricky.

First, you need to identify the target metrics or results you need to demonstrate to raise the next round. How do you do this? You can find some data online through case studies, founder interviews, or public data, but often you will need to set aside a couple weeks to talk to people who know the competitors and can tell you benchmarks for traction, like rate of growth over a period of time, number of users, or a key metric like session time or average shopping cart purchase size. How many users or much revenue did a competitor in an adjacent, comparable space have when they raised their Series A? This is your job to investigate.

Investors may ask you, "Why do you need to raise this much money?"

The most common answer goes something like, "Because I need to hire a designer and an engineer, and they will cost this much for this many months." But the better answer will likely be framed something like this: "This few million dollars will allow us to hit these specific targets, which we have identified will put us in a strong position to raise our next round, enabling us to hire these key positions as well as run key experiments and answer these open questions with the business model that we are building."

If you can nail this, you'll already be more prepared than most.

Here is a very rough, generalized process around the mechanics of fundraising to give you a sense of how it works:

Commit to a time frame for your fundraising process, and try to compress your meetings into that time frame. Treat it like a full-time job. During the fundraising process, you want to be meeting with people and pitching every day. At 500 Startups, we used to tell people that it takes 100 meetings for every million dollars you want to raise. The more networked you are as a founder, the fewer meetings needed, but this is a reasonable expectation.

Create your pitch. Do not open up Keynote or Power-point! The first thing you must be able to do as a founder is speak for sixty seconds about your idea and why it needs to exist. Find the three most cynical people you know, and tell them your idea. Find the three most positive people you know, and tell them your idea. Create your pitch deck slides. Show another five to ten people you trust who can help you iterate it.

Create a list of the founders you would most love to meet. Use LinkedIn and your current network to get connected. If all else fails, try to cold email them—my best tip: fun subject lines! Don't assume that because you're just starting out that they'll never speak to you. I have made vital connections, recruited great talent, and closed major deals from personal, short cold emails. The emails are the first step, but the key to gaining high-quality introductions is building relationships, which takes time. Don't expect someone you gave a business card at a meetup to open up their rolodex. It takes many weeks, coffees, and strong impressions to gain a founder's trust and earn an introduction. Learn about "double opt-in" introduction etiquette, and use it.

Make a list of everyone you know. Separate that list into three groups: tier 1 (your dream investors), tier 2 (solid, smart investors, perhaps not in your ideal space), and tier 3 investors (investors who are not your ideal), and start practicing on tier 3 investors to hone your pitch. After every meeting, take notes about the pitch, objections, and areas where your pitch was weak. Update your pitch deck. That way, when you nail your tier 1 meetings, your pitch has been sufficiently beaten up.

Figure out what investing vehicle is best for you—SAFE, convertible note, or priced round. Google these terms and know them. Get a lawyer who understands early stage venture capital. If you're very lucky, you may find one who is passionate about the space and will defer their fees.

Priced rounds are more complex, require multiple meetings, and more legal back and forth. Angel rounds are much more simple, and you can close money in a few coffee dates, since the angel investor is the only decision-maker. In these situations, know your closing question: "Can I count you in for $X?" was mine. Have the paperwork ready to send if they say yes. Send immediately, along with wire instructions to your bank account. Rinse, and repeat.

Raising funds is an important step, but it's just a means to an end. It's not an indicator of your value or success. All it takes is one yes to put a round together—every no brings you closer to that yes.

Next up, we'll explore risk and rejection, two essential pieces of fundraising.

HOW TO UNDERSTAND SEXTECH RISK

As a founder, you will need to understand your business model and your company's specific risk profile. In a blog post I read early on by Hunter Walk, an early stage investor at Homebrew and former product leader at YouTube and Google, I began to think about startups, especially my own, in terms of risk.[83]

Let's think about the main risks that investors care about when evaluating a startup: team risk, technical

---

83   Hunter Walk, "A Question About Risk That Founders Forget to Ask VCs," March 2017.

risk, scaling risk, market risk, growth risk, and financial risk—especially when it comes to sextech. While team, technical, market, and scaling risk apply roughly the same for sextech as they do for other startups, the central risks that sextech companies need to be particularly adept at addressing are growth risk and financing risk.

Team risk is the most important factor in any investment. Most VCs report that the founders are the most important factor in their decision to invest. The better the founding team, the lower the team risk. First time founders are sometimes treated as having a higher team risk. Founders with proven track records have a lower team risk. A good founding team is made up of people with highly competitive expertise, different lived experiences, complementary skillsets, and a leader with a clear, massive vision.

Technical risk comes with building a product whose success is based on the development of a new, innovative technology. If the company's technology fails, the product fails. This is not typically a web product or an app—this is the type of technology that requires a patent and likely a team of PhD scientists. Creating a teledildonic device that operates within a virtual reality world? That could be a higher technical risk. These companies are rarer.

Scaling risk is about looking into the future. The business

model might work now, but can it grow and expand—can it scale up? The bigger you grow, do the unit economics still make sense? Scale is important for VCs because they are only considered successful if they can achieve massive returns on their investment. That's why you rarely see a VC getting involved with consulting or traditional retail. Unless it's a business model that can really scale—say, become a massive franchise or disrupt an entire marketplace—it probably can't deliver the higher margins VCs crave.

Software and marketplaces are popular among VCs because they are particularly scalable. Take communications unicorn Whatsapp for example—when they were acquired, they were serving 450 million users, even though their team only had 35 engineers.[84] They had scaled at a phenomenal rate, and while many people in the United States hadn't even heard of the company, the potential for growth was incredible.

Market risk refers to the size of the market your business can address. How big is the market you propose to control, and how much of it is realistically available? As a sextech startup, market risk is where we excel. Sex is huge—current estimates suggest that we'll reach $122B

---

84   Cade Metz, "Why Whatsapp Only Needs 50 Engineers for Its 900m Users," *Wired*,
      September 2015.

by 2026. Every single adult human being on the planet is a potential customer.

Despite this, all sextech founders have to do some level of education to help their potential investors see the opportunity, and it's important to understand the total addressable market that you're going after. It's one thing to say there are 8 billion potential customers, and another to lay out just which among them you'll be targeting.

Airbnb's founders were laughed out of investor offices because most investors didn't believe there was a market for people staying in other peoples' houses.[85] Those investors were seeing the market risk—an industry already captured by global hotel brands. And they could have been right. But for those investors that Airbnb was able to convince, who understood the potential market the same way Airbnb's founders did, the rewards for that risk were enormous.

If the problem you are solving in sextech is very limited or for a very specific population, then VCs may see your startup as having high market risk. Be prepared to address, once you serve the specific population, how it can expand beyond the initial market.

---

85 Biz Carson, "Old Unicorn, New Tricks: Airbnb Has A Sky-High Valuation. Here's Its Audacious Plan To Earn It," *Forbes*, October 2018.

So, those were fairly easy. On to the hard ones.

Growth risk is the risk that you won't be able to acquire new customers at the rate you need. All consumer companies struggle with growth risk. Companies that can't sustainably acquire new customers die or stagnate. In traditional consumer companies, it's a fairly simple equation to try and determine acquisition—are you spending more to get them than you're making from them?

In sextech, we have a specific growth risk because we don't have the same access to standard paid advertising channels like Facebook or Instagram. Our referral and organic social media sharing activity is often hindered by stigma—meaning people won't often recommend their favorite vibrator publicly on their social media profiles the same way they might a meditation app. In addition, major media platforms are limited by their advertisers, so sexual wellness content is often subject to unfair restrictions there as well. We'll talk a lot more about this later—as well as strategies for overcoming it.

Financing risk is the biggest risk for sextech ventures—it means *will you have enough access to future funding to successfully grow?* Think about it like a tree planted in a small container pot. That tree could become a huge oak—but inside a container, it'll never be able to access the resources it needs to grow. Will you be able to get downstream capital for the business? (Downstream capital refers to capital that is raised at later stages of a company's life from larger, more institutional investors.) As importantly, what are the possible liquidity or exit strategies for your business? This is a way of asking: who could possibly acquire you? How will your investors achieve their goals and return their investment?

Smart founders will go into investor meetings knowing how to talk about these potential outcomes. The good news about financing risk is that investors are always jumping on the next big trends. One day it's software as a service. One day it's cryptocurrency. One day it's cannabis. Sexual wellness is poised to meet the needs of the next generation, and smart investors are beginning to recognize the opportunity. That's not accidental—it comes from the thousands of hours sextech founders have put in educating investors on this opportunity. My prediction is that, like femtech, which suddenly jumped to a billion dollars of investment in 2016, sextech will be an even bigger opportunity.

When someone asks you as a founder, "What are the good reasons an investor will NOT invest in you?" you should be able to answer them. All startups come with risks, and it's important to understand how others will evaluate your company.

HOW TO DEAL WITH REJECTION

When you're fundraising for sextech specifically, it's hard to know why someone has passed on investing in your company—they won't always tell you. Even if they love it, I've found VCs are more likely to make up reasons why they don't want to invest in you rather than admit the real reason.

Maybe they are not comfortable bringing it up in a partner meeting with their colleagues. Maybe they know they have a conservative LP for whom the investment would raise their eyebrows. Maybe they are struggling to talk about sexuality because they are ashamed about it themselves. Maybe they actually don't believe in the space. Maybe you have not made enough progress or gathered enough data. Maybe you haven't actually figured out an interesting business model or customer acquisition strategy—maybe they are right not to invest. Maybe they just don't believe in you. All are highly possible. None of these should dissuade you from keeping on your path.

Venture capitalists have the same issues with sex as the larger culture—and, unfortunately, no one puts how sex-positive they are on their Angelist profile. You have to find it out yourself. Maybe you find out the hard way when they shut you down with patronizing or ignorant talk. Or maybe you find out later from people who know them well. One of the ways I learned whom I should approach was by listening to what other people were saying while at parties, hanging out, and networking. I learned to ask people, "Who do you know in your investor friend group who is pretty open-minded?" That's where I'd get my referrals—and it's where I'd later hear about how things really went.

While launching O.school, I had an early meeting with a major investor—our initial meeting was great. He had a million intelligent questions to ask. We went through the standard phase of oversharing. (I used to joke about putting a slide in my deck that said, "This is the part where you can overshare with me about your life.")

I couldn't get an answer from him as to whether he would invest, but the signs were good. At parties people kept telling me how he couldn't stop talking about O.school. He loved what we were doing and talked about how important it was. He was almost as big an advocate for the company as I was—at the same time, he wasn't answering my emails. He didn't return my calls.

After months, I got an email with a generic shutdown: "It's a no from me."

I wouldn't find it out through him, but through mutual connections I learned that his fund had a particularly conservative Limited Partner. As much as he loved the product, and as much as he wanted it to succeed, and as much as he believed it would, he couldn't do it himself.

(On the opposite side, I've had meetings with VCs at funds who have refused to invest from the fund—but then invested personally. You never know!)

Once—again through friends—I heard an explanation from a well-known woman investor: "I love that O.school exists for my daughters, but it's tough for me to put my name on this. I don't want my reputation to be that I'm a woman who invests in sextech."

Sometimes it's just not about you. You dust off, and keep going.

My advice: build a strong community of other founders, advisors, and supporters around you. Find people who can backchannel for you if you need them to call up a mutual connection, and also people who are just there to listen after the harder days. They are your shield, your sanity squad, your reminder that you are not alone and that people believe in you, even when you're finding it hard to believe in yourself.

HOW TO GET A BANK ACCOUNT

When I was getting started, people warned me that banks

would be an obstacle for a business centered around sexuality. At that point, though, I'd worked in accounting for years. I'd worked with every major bank through inDinero, both in Silicon Valley and nationally. I knew the system well and believed that I could get around it.

Like the word of mouth referrals I'd get later for investors who were "open-minded," people started telling me which banks were too conservative and which banks might be okay. A few people with great reputations recommended a premier Silicon Valley bank where I had many connections, so I set up a conversation and explained what we were doing with O.school. I let them know what the company was about and that I was starting to fundraise. In a conversation early on, I was clear with them: "I'm about to raise some money. If the sexual nature of our business is going to be an issue for you or your risk department, just tell me now. I've heard there are going to be problems, and I want to know sooner rather than later."

They loved what I was doing, they loved my background. They said they'd talk to their risk department, but that we weren't going to have any problems. Amazing.

O.school barely had a landing page at that point. There was nothing sexually suggestive whatsoever—just a stock photo and an email capture form. For all intents and purposes, O.school wasn't even a thing yet. Opening the bank account went smoothly. I raised money and it went into the account.

About midway through fundraising, I got a short email from the bank. They were asking me to take my money and leave.

Even though I'd asked them to be sure first and to not lead me on, once the risk department got hold of our account, they freaked out.

That was my first "Oh, shit" moment. We hadn't transacted any business, we hadn't even written a single line of code. We hadn't done anything even remotely related to sex or sexuality. None of that mattered.

My network was stunned. It was a humbling moment for me, because that day I realized my network couldn't help me. If I, with my extensive finance and tech background, couldn't get past the bank's risk department, what chance did other sextech founders have? That's when we learned to get creative.

PICK YOUR BATTLES

Leveraging connections can help, but it's not your only option. When I ran into banking problems, the suggestion everyone had for me was to go to an "adult bank" that charged high fees in order to offset perceived risk.

"Don't make a fuss about the banking situation," I was told. "If you try to take a stand, write a Medium post for every injustice, and fight all of the battles at once—you will waste time. Your time and energy are the most valuable assets of the company. Your job is to make your business successful, not to fight every fight that comes your way." It was hard to hear in the early days, when I was so mission-driven. It's a hard balance to strike when you're facing blind, stupid injustice. There are times when I've spoken up about big systemic issues, like Apple's parental controls policy that banned us. Other times, the risk to your company is just too high.

I couldn't fight the whole banking system—I just wanted to find a way to grow my company.

So this is my advice to other sextech founders: Don't go to the boutique banks that offer custom services to big tech companies. Go to a bank where you're not special at all. Where to them, you're just a company, and one of many. Hide in plain sight. Rather than go to the bank in the Financial District where all the startups go, and where risk profile might be scrutinized more, I simply went to the local branch of a big, national bank far from the Financial District and closer to the Tenderloin of San

Francisco. I told them I needed to open an account for a software education company. That's it.

In the two years that I've been running O.school, I've never had a problem. No one is looking hard at what we're doing, and no one has had a problem with us.

When looking for banking, don't offer more information than what they need to open the account. Keep it vague. Get what you need, and get out. Resist telling them how great your sexual wellness company is going to be. While it might not seem as exciting as picketing outside the bank, it's actually more radical than it seems—the more embedded we become, the more we're able to point to successful sextech companies at major banks with no greater risk, the easier it will be for future companies to get a bank account.

HOW TO MOVE MONEY

To accept online payments, you'll need a payment processor.

Processors are the intermediaries between you and the bank—and they all are governed by rules set by Visa/MasterCard. Some of these are fairly clear, like no selling drugs or illegal products. But some are a bit more vague—especially when it comes to sex. If at any point they don't like something you do, like using words on their blacklist or not wording your community guidelines conservatively, they will put you on a blacklist. No one will be able to process for you, meaning you won't be able to take payments. At the very least, you'll have to re-incorporate your company. Sometimes, even that won't work. It's a very serious problem.

The other option is a high-risk payment processor—processors who are used to dealing with a high-risk customer base. They know there aren't other options for their customers, so they are often more expensive. It's like a sin tax—6 to 10 percent processing fees compared to an average of 2 to 2.5 percent for everyone else.[86]

When O.school was implementing the tip system for our live stream product, I anticipated that we would need payment processing. Most modern tech companies rely on companies like Stripe, a multibillion-dollar company whose whole job is to provide the payment processing functionality for other companies. It can save weeks of development time and resources to implement a plug-and-play, off-the-shelf tool like Stripe, so it was really important to me for us to be able to partner with Stripe for O.school.

---

86 Motile, "Adult Merchant Accounts: Professional Services Tailored to Your Business," Accessed online, June 2019.

In my first company, inDinero, we relied on Stripe, and we were actually in their first batch of customers (at least in their first thirty or so) back when they were called DevPayments in 2010. Seven years later in 2017, I called John Collison, one of the founders of Stripe, to ask him to invest in O.school. He politely declined the offer to invest, which I assumed he would, as he and I had barely interacted. But, next, I asked him if he would help us get a Stripe account, which he enthusiastically agreed to. He directly emailed someone internally, and a few months later, we had a Stripe account and a dedicated person we could email if we had any problems. This was a great turn of events, but it's really unfair that a personal favor and chance relationship to one of the founders of the biggest payment processors was what it took to help us get a payment processor account seamlessly. I've spoken to quite a few sextech founders who have struggled with payment processing, and tried to help where possible. The harder companies are the adult brands that include nudity, where the line is very strictly high risk, whereas O.school and more wellness and health-focused brands are beginning to break through. I hope that one day more banks and payment processors create solutions for all companies, including ones with nudity.

Unfortunately, many of the standards are decided on a completely subjective basis. No one will tell you who makes these decisions, or the rubric for abiding by them.

The clauses, they're very ambiguous, and there's no appeals process. All decisions are made by someone's boss's boss—and you can never actually reach or speak to them.

When I was researching payment processors, I was told about another potential danger. A lot of payment processors get big by working with everybody, no matter what they do—in most cases, meaning they're porn-related. The processor will grow their business by working with adult businesses, raising more and more capital—until they reach a certain point. At that point, in the downstream stage, they'll look at the risk profiles of everyone with an account and fire all the companies at a certain risk level.

Worse, a conservative LP somewhere in the organization will speak up, and you'll suddenly be kicked out. It's unlikely that anyone will stand up on your behalf, because few individuals have enough leverage to change these decisions once they are made.

Even Cyan Banister, a tech pioneer who had deep connections to the PayPal network through her partner Scott Banister, an early advisor and board member at Paypal, didn't have enough leverage. In 2007, Cyan launched Zivity, a premium subscription and fan interaction site devoted to nude photography. To fight against stigma, Cyan even posed for photos herself. She was forward-thinking, on the side of free speech, and fighting the battle to allow people to celebrate this type of beauty.

By August 2007, she had raised a million dollars in seed funding and was growing an incredible community. And PayPal shut her down...twice.

She wouldn't go down quietly. Cyan told me: "If you believe in the space and want to build something in the space, you have to be willing to fight for it. A lot of people said 'Well, you're an adult company, what did you expect?' That made me angry, and I channeled that anger into finding solutions. I didn't put all my eggs into one basket and instead found multiple processors."

Today, she's a VC at Founders Fund with an incredible profile including Uber, Niantic, Postmates, DeepMind, Affirm, and SpaceX. She's as connected and respected as someone in Silicon Valley gets—and had the same issues. Don't take it personally.

The origins of sextech and porn being "high risk" is a vestigial practice left over from the early days of the internet, from a phenomenon called "friendly fraud." Adult content was one of the first products that could lure a consumer to enter a credit card into a website.

A husband would buy porn, but the wife would get the bank statement. She'd ask husband, "What's this charge for $69.99?" and ashamed, he'd play dumb, "I don't know...wasn't me! It must have been fraud." The wife would call the payment company and tell them it was a fraudulent charge, which would result in a "chargeback"— the money would be refunded by the porn company.

Each chargeback would result in a "ding" on the company's record. The more chargebacks a company receives, the more likely they would be designated as a "high-risk" company. Adult companies used to have really high chargebacks thanks to friendly fraud, impulse purchases, and shame.

Today, it's completely outdated. Credit card processing is much more discreet, and the taboos are slowly beginning to lessen. Yet this old, outdated phenomenon is being used to determine risk profiles. From what I've gathered, sexual wellness companies have the same chargeback ratio as any other startup.

Stripe is a payment processor that is slowly beginning to back sextech, and I recommend trying them.

HOW TO FIGURE OUT CUSTOMER ACQUISITION

The most common reason a consumer company dies or

stagnates is when it struggles to acquire customers sustainably. Whether you are building a product or spreading content or building a community, your customer acquisition strategy is the single biggest indicator of whether you will be successful.

If you are marketing a product, your main channels will be content marketing, paid marketing, word of mouth marketing, and public relations (PR), like press mentions and broadcast appearances.

Press mentions can have an impact, but it's rarely the only avenue you can rely on. Good PR can definitely get companies off the ground, and it can bring a lot of attention, but it's not a predictable, sustainable way to grow. You can't really promise an investor that you'll have a viral hit every month. Also, driving traffic from a PR mention to your site can be hit or miss. Some of our best PR hits came from mid-level publications while an op-ed in the *New York Times* may only garner a few thousand visitors. You never know. PR can be a powerful brand builder and traffic driver, but it's rarely the key or central strategy.

The most common way that sextech companies acquire their customers is through creating high quality content for their community, leveraging social media followings, building strong newsletters, and other creative ways to engage, educate, and reach their desired customers directly. It is not easy to create high quality content, but it's a high impact strategy, especially at first, to build a customer base and learn. I suggest most companies start there to test brand messaging and get clear on the audience needs and personas.

Next, you may try to distribute that content through influencer partnerships, or even larger installations like the MTA, billboards, and other types of paid advertising placements. If you have amazing content and a strong brand, these are experiments you may want to try. The key is experimentation, which requires discipline with tracking, analytics, and considerable strategy and execution.

The downside to content creation, other than the cost associated and the more vague ROI on cost per content production, is that it can take more time than, say, paid advertising, the channels that modern sex toy companies like Unbound and Dame Products have been fighting for. Let's dive into how to navigate paid advertising.

HOW TO NAVIGATE PAID ADVERTISING CHALLENGES

To start thinking about a paid marketing strategy, it's important to understand unit economics, such as the ratio of the cost to acquire a customer to the lifetime value of a customer (sometimes shown as CAC:LTV). Say you have a condom subscription company—you can figure out how much each customer will spend each year, about how many years they'll stay with you, and that's your lifetime value (make sure to account for 'churn,' or the rate at which customers unsubscribe if it is a subscription business). The higher the lifetime value, the more you can spend on customer acquisition.

For e-commerce companies, it may be the average cart size of an order and number of repeat orders in a given time period. For young startups, it takes a few months of data to know exactly, which is why it's important to have friends in similar or adjacent spaces who can offer this type of data as a benchmark.

Paid marketing can be a useful, critical customer acquisition strategy for user growth. Over 70 percent of all advertising spending is happening in Google and Facebook right now, which has huge repercussions for media and other advertising-supported businesses.[87] Facebook and Google ads are important because they are cost-efficient. The consequences of not being able to use them can be frustrating, as companies want sustainable growth to be able to secure investment. Many direct to consumer brands scale up using these platforms, because they work.

---

87 Jillian D'Onfro, "Google and Facebook extend their lead in online ads, and that's reason for investors to be cautious," CNBC, December 2017.

You will likely not be able to advertise your sexual wellness business the same way that other companies do unless you are extremely strategic about ad copy and images.

Even with your best efforts, your Facebook ads will likely be disapproved at first. Instagram pages may be taken down without notice. Twitter, Pinterest, and Reddit may ban you from paying for advertising. And it's gendered, just like the MTA—Viagra is fine, but the word clitoris is not. Even breast pump companies get their ads disapproved on Facebook and other platforms because of blanket bans related to showing skin in an ad with the word breast.

Alex at Dame Products figured out she could boost articles about herself as the founder of Dame Products but not her sex toy company directly, so she started writing more content and using that strategy to drive traffic. Toy companies have a particularly hard time, because Facebook seems to have a pretty strict view on vibrators and other sexual products.

Initially, O.school also struggled. When we tried to experiment with driving traffic, 95 percent of our ads were disapproved on Facebook. We were told we could not advertise on Reddit unless it was a NSFW (not safe for work) channel, which was not the audience we wanted to reach. So, growing our organic reach was all about producing high quality content.

Creating valuable content is great because it lasts, and ultimately creates more customers while you do it. People who are searching for "the magic wand vibrator" already know they want to buy it, so the competition for the paid ads around those "high intent" search terms are more expensive and competitive. You could compete with everyone else who is selling a magic wand—or you could educate the people who don't even know what they want yet.

HOW TO THINK ABOUT DARK SOCIAL

Word-of-mouth is amazing, and the lifeblood of many companies. If you can grow purely by word-of-mouth, do it. You could build tools that facilitate more sharing through word of mouth. In any market, it's important to build something that people love so much that they do the marketing for you. But when it comes to sextech or sexual wellness, word-of-mouth is more often a whisper. Will people behave the same as they do with other products? Will they talk about it and share it with their friends? Not necessarily.

If you're the type of person to start up a sextech company, it's easy to forget that most people in the world don't share your mindset. You might proudly share your work on social media, but your analytics and marketing analysis likely will tell you that most people won't feel comfortable enough to publicly talk about your product. They're not going to support a vibrator in the same way they might support a new piece of luggage. They're worried about their work friends or mother-in-law seeing it.

Dark social is a phenomenon in which people don't use their public platforms to share. They might email it or text it, but it won't go viral publicly.

Dark social is difficult to track. At O.school, we often don't know how people come to our sites, which is why we call it "dark." Alexis Madrigal actually coined the term in the Atlantic in 2012. We get traffic from people sharing things with each other privately rather than from trackable sources. This may show up in a pesky "organic direct search" category on Google Analytics, which shows up as direct searches for your company's name. Attribution is getting better every day, but this is still a very hard source of traffic to track.

As sexual wellness becomes more normalized, this may change. In the meantime, you have to accept that marketing and metrics don't work the same in sextech as they do in most other markets.

That doesn't mean we can't learn from other sectors. I learned how to manage this by studying with other finance-related brands, since they also deal with shame. After all, no one wants to share an article about bankruptcy, either. Many of those companies explore private messaging, chatbots, community forums, SEO—making sure you show up in Google searches—and other channels that respect someone's need for privacy.

HOW GATEKEEPERS MAKE US STRONGER

The ocean might be blue, but that doesn't mean it's easy to build a sextech business. The usual avenues may or may not work for sexual wellness entrepreneurs. But the gatekeepers' power isn't the same as invincibility. Sextech entrepreneurs don't hold back just because someone in a traditional platform says no. Instead, a gatekeeper's refusal to engage with sextech rings a sign of how they will become irrelevant if they do not adapt. Rather than a sign of power, it's a weakness that can be used against them.

Building a business in spite of systemic hurdles tends to shape a stronger, more resilient platform. Defunding and de-platforming are constant threats for the sextech entrepreneur—which means we can't get comfortable. We have to build innovation and adaptability into the DNA of our businesses.

Cyan Banister put it this way: "When you're a founder, it's hard to be excluded from these institutions that represent capital and success. It's unfair to have to work harder, and fight for the access and accolades that seem to come so easily to those who play by the rules. But the work we do makes us stronger companies and better leaders. It's like high-altitude running—and I'm firmly convinced that the leaders in sextech will one day be recognized not only as pioneers, not only as risk-takers, but as the most agile, determined, and successful entrepreneurs in tech."

HOW TO DEAL WITH LANGUAGE

We are in the middle of a time of great experimentation, not only technologically but socially. We've torn down some of the old rules of gender and sexuality—and we're in a mad, creative, exciting scramble to draft the new ones. No matter what you say, you're likely to be criticized, sometimes harshly—often by people you respect. Use that and learn.

You'll come to realize through criticism that sometimes your dream didn't take into account everyone. It's also likely that, at this early stage of your business, you're not able to do what you know needs to be done. Find a way to absorb criticism, and understand perspectives, without giving up.

Of course, as you're fighting to change the world, you'll get backlash from the institutions, the people that still hold the majority of the power. They're still holding on to fear and shame.

In creating a practical product, you may sometimes feel alone—caught between the ideals of activism and the demands of the institutions. You'll have to keep your balance between those two at every point within your journey. Nobody is perfect.

For example, the language that we use in some "woke" spaces won't be the language your average consumer uses. We have to find a way to use language people will understand, while still making sure we're not excluding marginalized communities. Part of your duty as a sextech entrepreneur is to use less esoteric language to communicate broader ideals. A good place to start:

- Opting for "reproductive health" instead of "women's" or "men's" health
- Referring to body parts instead of gender to be trans-inclusive like "vulva owners" or "penis owners" instead of "women" and "men"
- Committing to sharing pronouns, and respecting they/them pronoun use, making it part of culture to talk about pronouns, even before any gender non-conforming person has joined

At least at first, you'll need to identify your market and use language that meets them where they are. But make sure to never lose sight of your ultimate goal—that vision will allow you to expand your impact once you're on more stable ground.

CALLING OURSELVES IN

I believe wholeheartedly in the power of education, and because of that, I choose to believe that people can change. One of my core guiding values from Maya Angelou is that "when people know better, they do better."

You may have noticed that throughout the book, I chose not to name people who weren't perfect. I didn't name the women I spoke with or any of the investors or founders I mentioned in the book, because I think that public shaming, when used injudiciously, can harm rather than help. We deal with enough shame in the world, and I'd rather not add to it if I can avoid it. As a Filipina, I was brought up to believe in loyalty and community. As a result, I try to give constructive or critical feedback in private and praise in public. It means that in my communities, from my pinay community to the entrepreneur community to the sextech community, I try to call people and connect directly, and give people the benefit of the doubt when misunderstandings and mistakes happen.

As we work together to build a new ecosystem, I'd like to see some grace built into the sexual wellness community.

Too often, I think, public shaming flattens the conversation. It resorts to clear binaries—us versus them, wrong versus right—that can limit our ability to move forward. Public shaming is the tool of the patriarchy, white supremacy, and many religious institutions, and it is not a tool that communities should use against one another.

O.school's initial model was live streaming video. In the first months after launch, we were thrilled to see an incredibly diverse group of voices go online and begin educating. It was tremendous, it was experimental, a free-for-all of sexuality education. Fewer than six months after we launched, President Trump signed a law that made platforms legally responsible for the content on them, specifically when it came to sex work, as a way to address sex trafficking. The law was written so broadly, and so badly, that no one—not the activists that opposed it, not the legislators that passed it—really knew what it would mean in practice.

It was called SESTA/FOSTA, and was so vaguely written and harsh that sites like Craigslist immediately started shutting down entire sections of their platforms, lest they be held legally responsible for posts by sex workers.

At that point, O.school had welcomed open discussion of sex work. In fact, in the week after it was passed, one of our educators had a class scheduled specifically on sex work. But suddenly, there was the possibility—however low—that any member of my team could face serious retribution or at least the platform could face consequences if we were accused of violating the law. It wasn't entirely clear, but when I saw Cloudflare—a large company with over $150M in funding and a commitment to free speech—report they would be complying with the law, I called our lawyer.

Before meeting our lawyer, we asked the educator to postpone the class, or shift the content, while we figured out what was going on. I didn't see this as a permanent change, but a temporary pause while I made sure that we were in the clear.

The educator, who was a proud and respected sex worker, was not happy, and I don't blame them. At a time when their community was under attack, one of their allies appeared to be buckling. Across the country, platforms that served sex workers were being shut down, their ability to generate income destroyed, and their ability to communicate with each other decimated. Not only was SESTA/FOSTA a massive free speech violation, it would more importantly put sex workers' lives in danger. They were outraged, and so was I.

Had we been able to speak, I might have explained all this to them. That the survival of the entire platform required time for the lawyer to understand the new law—something few lawyers outside the sex work community had even heard of. But as deserved as their outrage was, I felt it was my responsibility to take SESTA/FOSTA seriously.

Within a day, a blog post went up accusing O.school of turning its back on sex workers, among a list of other grievances, all of which I would have welcomed as feedback and taken swift actions to address. A few educators on the platform quit in solidarity, and I began receiving calls from tech press. I was happy, in a way, to speak with a journalist because, while it wasn't positive news, I was able to explain all the things that I would have explained had they reached out.

But that's the problem with call out culture. Even though people had direct access to me, not one of the educators had called me before blasting their posts. Not one had even texted me or sent a direct Slack message. The cycle went straight to public outrage—tweets, blog posts with images of my face, and ultimatums. I understand the rage, but in the end, it wasn't productive.

The people who did call me were other women of color who called me to say, "Hey, we know you're trying your best." It was one of the hardest moments for me as a founder, but it also taught me so much about grace. I'm grateful for the experience now.

The internet has made speech so much more democratic, which is a wonderful thing. But social media also silos us, and allows us to react publicly before connecting privately. I ask that as we all move forward, we allow each other grace. That we call one another directly when something goes wrong. That we work together, without tearing each other down. We're fighting for power that has been denied to us for too long. If we don't build these things together, we'll get cut down before we've had a chance to grow.

Community is an ongoing relationship of support. It's not about jumping in and expecting everyone else to keep you on track. If you're starting out, you better spend some time actively learning from people in the sexual wellness industry, especially women of color, LGBTQ people—people with different lived experiences than the ones you've lived. They've been doing this work longer than you, and against greater odds. Work with them, support them, and listen more than talk.

Have a self-care plan in place. Hang out with friends and family who know you and can remind you of who you are. And please, if you can afford it, go to therapy. A lot. You're probably going to need it.

We're trying to change the world and build great businesses. That's not easy work. Criticism is going to come up. You're going to get stuff wrong, and you're going to hurt people and be hurt. Take care of yourself, take care of each other, and build resilience that can carry you through. Know that this is part of our journey toward a new future, and let that journey hold you accountable to your values.

Invest in things that build toward humanity. Educate everyone everywhere. Embrace the changing tides in sexuality and gender. Then join me—help the world, make some money, shape the future.

Be open to what's to come. Take time to let your vision unfold. Be open to learning and growing and constantly healing along the way.

Changing the world is exhausting. You know what builds energy and strength? Orgasms. They are one real fuel of the resistance. Don't forget yourself or your own pleasure.

Imagine a future that's built for healthy, whole humans— and then go build it.

No shame allowed.

ACKNOWLEDGMENTS

This book would not have been possible without the help of many generous minds and hearts.

Thank you, Brannan, Mike, Kelly, Navy, Justin, Gina, and Betty for reading drafts and patiently guiding me through my first book experience. Thank you, Jess, for predicting this book into existence, pushing me as you always do. Thank you to the whole O.school team and community for building this movement with me, and to our investors and supporters who have our backs. Thank you to my many friends who hold me up and always pick up the phone for me. Thank you to Arina, Sharon, and all of my mentors and pillars of support.

Thank you to the sextech and sexual wellness community for our shared missions, scars, and dreams. I love sailing around this blue ocean with all of you.

Lastly, thank you to my family, especially my amazing mother and father, Justin, and Celine, for always reminding me where I come from and where I'll always find home. You are the loves of my life.

ABOUT THE AUTHOR

ANDREA BARRICA is the CEO and founder of O.school, a judgment-free resource to learn about sexuality and pleasure. She's one of the only queer women of color to raise millions of venture capital for her sextech company. Previously, Andrea co-founded the leading financial solution for growing startups, inDinero.com, and served as a venture partner at 500 Startups, where she invested in startups all over the world. She overcame fear-based sex education in public schools and a strict Filipino-Catholic upbringing to become an in-demand sexual wellness speaker and a leading sexual wellness industry contributor to Forbes.com.